THE COLD-BLOODED AUSTRALIANS

THE COLD-BLOODED AUSTRALIANS

Gunther Schmida

DOUBLEDAY
SYDNEY & AUCKLAND

To Amber, Caprice and Holly

First published in Australasia in 1985 by
Doubleday Australia Pty Limited
91 Mars Road, Lane Cove, NSW 2066

National Library of Australia
Cataloguing in Publication data

Schmida, Gunther.
The Cold-Blooded Australians.

Bibliography.
Includes index.
ISBN 0 86824 111 3.

1. Reptiles — Australia. 2. Amphibians — Australia. I. Title.

597.9'0994

Typeset in Australia by Savage Type Pty Ltd, Brisbane
Printed in Japan by Dai Nippon

FOREWORD

Since I read my first book about Australia when I was thirteen, I have been fascinated by its unique wildlife — particularly its reptiles and frogs.

When I arrived in Australia in 1965, I spent a lot of time travelling around the continent, taking photographs of these animals whenever I had the opportunity. Despite a childhood interest, I virtually ignored the Australian species of freshwater fish during this time, and it was not until I finally settled in Sydney that I turned my attention to them again.

The idea for a photographic study of the three groups of 'cold-blooded' or ectodermic Australian animals was born several years ago. However, the pictures in this volume have been compiled over a period of seventeen years.

In this time interest in Australian fauna has grown steadily. Many new species have been discovered, and many interesting findings about the biology of some species have been made. Despite the literature available, our knowledge of the 'cold-blooded' fauna is still limited.

This book is not meant to be a comprehensive account of all the known Australian species, rather it is designed to show a *selection* of these freshwater fish, frogs and reptiles. I have tried to show them from their best possible angle, or captured in a unique demonstration of behaviour. Many of the pictures were chosen purely for their visual impact, and therefore an expert might consider that some animals should have been included. I have omitted some species for lack of a suitable photograph — for example, the taipan, Australia's largest and most dangerous snake. Although I have taken many pictures of this reptile, I have yet to take a shot that shows it at what I consider its best.

Where they exist, common names have been used in this book. The scientific names are as up-to-date as possible, although as continuing research uncovers relationships between species, name changes are frequent.

A lot of information given in *The Cold-Blooded Australians* has become common knowledge through the often unrewarded work of researchers. However, some is based on my own experience, and that of my colleagues. Likewise, many of the photographs have only been made available through my friends and through the generous supplies of specimens from all over Australia.

I am very much indebted to my friends John Cann, John Edwards and Steve Swanson, who taught me much about the ways of reptiles when I started, and for assisting me when taking pictures. For assistance in the field, allowing me to photograph specimens in their care, and for suppling fish and information I would very much like to thank the following:
Neil Armstrong, Vic.; Dr Gerald Allen, WA; Joe Bredl, SA; Bill Boustead, NT; Gary Backhouse, Vic.; Dirk van Beusekom, NSW; Rob and Julie Carroll, Qld; Dr Harold Cogger, NSW; Dr Jurgen Clasen, Germany; Neil Charles, Qld; Margit and Karen Cianelli, Qld; Dr Philip Cadwallader, Vic.; John and Kerrie Davies, WA; Keith Day, NT; Harry Ehmann, NSW; Graham Heidke, Qld; David Frusher, Vic.; Darryl Grey, NT; Bob Grey, NSW; Dr Allen Greer, NSW; Neville Hamlin, Qld; Greg Harold, WA; Paul Horner, NT; Brian Hancock, WA; Dr Walter Ivantsoff, NSW; Peter Krauss, Qld; Kim Kennerson, NSW; David Knowles, Qld; Glen Laycock, NSW; Ray Legget, Qld; Mike Mahoney, NSW; Keith Martin, NT; Brian and Glen McGregor, Qld; Greg Mengden, NSW; Dr John Merrick, NSW; Robert Pulverenti, NSW; John Rigby, NSW; Shirley Robinson, ACT; Gordon Stables, Qld; Iain Stewart, NZ; and Steve Wilson, Qld. To all others who helped but are not mentioned here, I also offer my thanks.

Thanks are also due to the designer Mike Blore for letting me interfere as much as I did, to my editor Peter Huck and to Lisa Highton and Doubleday for their enthusiasm and patience.

Last but not least, I thank my wife Angie and my daughters Amber, Caprice and Holly for their support and patience shown when I was travelling or tied up for hours trying to get that ever elusive 'final' shot.

Contents

INTRODUCTION

The vividly coloured **corroboree frog** (*Pseudophryne corroboree*) is named after a ceremonial Aboriginal dance in which the participants paint their bodies with vivid patterns that resemble the frog's markings. This amphibian is found above altitudes of 1200 m in the Australian Alps and grows up to 3 cm long.

ALMOST 1000 ANIMAL SPECIES, or over half of Australia's vertebrates, are cold-blooded. This figure includes some 190 species of freshwater or inland fish, 180 frogs and 600 reptiles. However, because new species are still being discovered and known ones reclassified these statistics should only be regarded as approximate. Instead of examining each class of animal as a separate entity this book tries to convey the incredible richness and diversity of cold-blooded animal life found throughout Australia by choosing representatives of each class and grouping them together under eight different drainage regions. As such the animals discussed in this book are only a selection and not a definitive list of all the cold-blooded creatures found in Australia. Of course every division has its drawbacks and it must be stressed that some species mentioned occur in more than one drainage region while others are only encountered within a small area or habitat in one region. For instance, the racehorse or Gould's goanna (Varanus gouldii) is widely distributed throughout seven regions and is also found less frequently in the eighth while the western swamp tortoise (Pseudemydura umbrina) is restricted to a few selective swamps near Perth in the South West region. Before looking at the typical occupants of each region this introduction is intended to provide a brief survey of each class of cold-blooded animals in turn, starting with the fish and continuing with the frogs and reptiles.

Compared to the 3000 or so types of marine fish found in Australian waters the country's total selection of freshwater and inland fish is rather meagre. Less than 130 species are known to spend their entire lives in freshwater while another 60 odd types frequent both freshwater and saltwater habitats. Out of this total only three families are regarded as true fluviatile fish in the sense that they evolved completely in freshwater. They are the bizarre Queensland lungfish (Neoceratodus forsteri), found in the Burnett and Mary River systems of central Queensland; the saratoga and spotted

WATER DIVISIONS OF AUSTRALIA

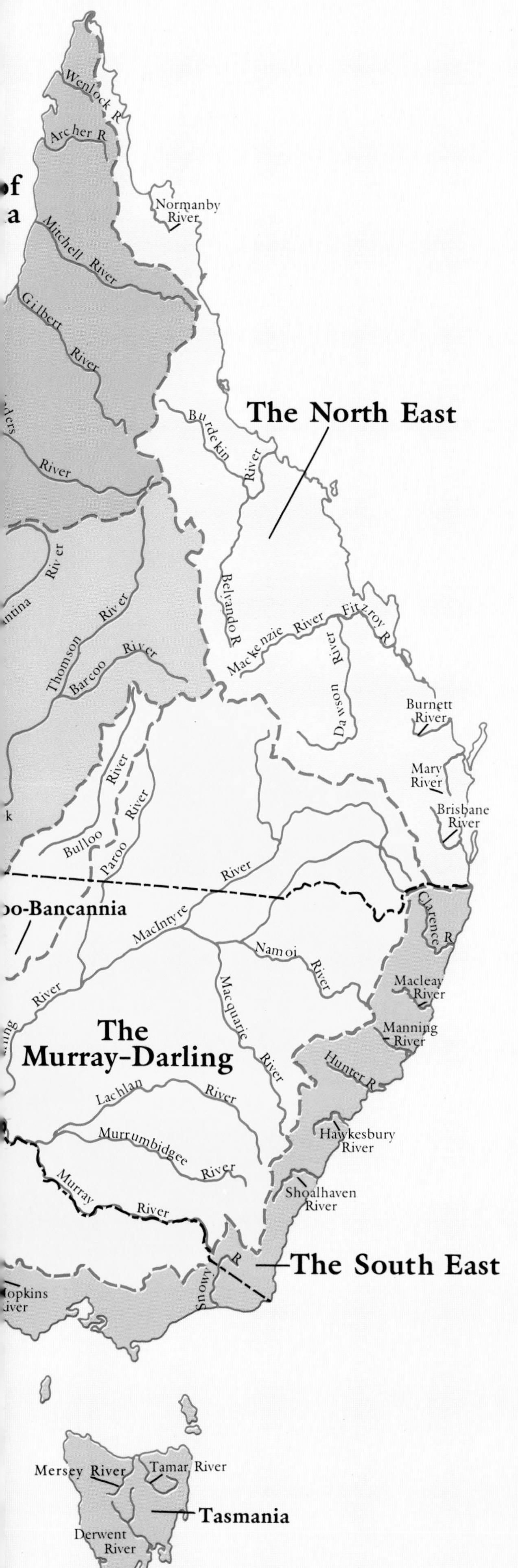

barramundi (family Osteoglossidae), distributed throughout Queensland's Fitzroy River system and selected waterways in the Northern Territory and Cape York Peninsula respectively; and the salamander fish (family Lepidogalaxiidae). Possible primary freshwater fish are also the galaxias or mountain trout (family Galaxiidae), smelt (family Retropinnidae) and grayling (family Prototroctidae), all of which inhabit cooler waters in the southern part of the country. The other freshwater fish found in Australia are, in a geological time sense, comparatively recent arrivals from the sea and still considered to be in the process of speciation. Little was known about Australia's fluviatile fish species until fairly recently and one of the consequences of more thorough investigation and research has been the recognition of several new species and the reclassification of others, erroneously regarded as new and since found to be variations of other known species. This process of discovery and reorganisation is likely to continue.

Thirty-nine families of fish are known to exist in Australia. Besides the species mentioned in the previous paragraph some of the better known fish are as follows. The most common fluviatile fish in Australia is undoubtedly the herring (family Clupeidae), a species rivalled numerically in some northern waters by rainbow fish and blue-eyes (family Melantaeniidae) and also by the hardyheads (family Atherinidae). Because of their abundance these fish all form an important element in the food chain. Well-known predators include the long tom (family Belonidae), distinguished by its pincer-like jaws and rows of razor-sharp teeth, and several perch-like species including Murray cod and golden perch (family Percichthydae), sooty and spangled grunters (family Teraponidae), jungle perch (family Kuhliidae) and silver barramundi (family Centropomidae). While the Murray cod and golden perch are found in the Murray-Darling River system the other fish frequent the northern river systems. Smaller perch-like fish include the Archer fish (family Toxotidae), the mouth almighty (family Apogonidae), the glass perch (family Ambassidae) and the pigmy perches (family Kuhliidae). Two

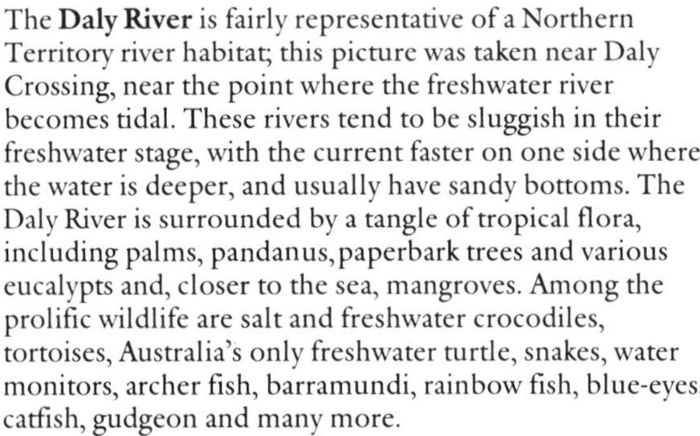

The **Daly River** is fairly representative of a Northern Territory river habitat; this picture was taken near Daly Crossing, near the point where the freshwater river becomes tidal. These rivers tend to be sluggish in their freshwater stage, with the current faster on one side where the water is deeper, and usually have sandy bottoms. The Daly River is surrounded by a tangle of tropical flora, including palms, pandanus, paperbark trees and various eucalypts and, closer to the sea, mangroves. Among the prolific wildlife are salt and freshwater crocodiles, tortoises, Australia's only freshwater turtle, snakes, water monitors, archer fish, barramundi, rainbow fish, blue-eyes, catfish, gudgeon and many more.

This typical **Snowy Mountain** habitat vividly emphasises that not all Australian conditions are uniformly hot and dry. During late winter and spring there is considerable run-off from melting snow and numerous fast flowing streams, such as this one, carry water from the high tussock country into the Murray-Darling system. These streams are characterised by low water temperatures and high oxygen levels and are the home of the mountain trout and other native species.

This **red sand desert** in central Australia is actually used as sheep grazing country, although the absence of low shrubs and plants indicates that it has been overgrazed. This type of landscape is most verdant after rain when a carpet of highly adapted plants emerge, briefly completing their life-cycles before dying. In between, grevilleas dominate the landscape. Despite the desert's desolate appearance it is the habitat of numerous cold-blooded animals, including some fifteen species of geckos, many different dragons, countless skinks, goannas, legless lizards, pythons, small burrowing snakes as well as large venomous species.

catfish families are found in Australia; the fox-tail catfish (family Ariidae), which lives in northern waters, and the eel-tail catfish (family Plotosidae), which lives just about anywhere with the exception of cold mountain streams. Gudgeons (family Eleotridae) are commonly found in Australia with one of the more unusual species from this family being the blind cave gudgeon (Milyaringa veritas). Other unusual species include the nursery fish (family Kurtidae), found in the coastal swamps and ditches of south west Australia, and the desert goby (family Gobiidae), one of the few fish types able to survive in the arid interior and restricted to small pools formed by artesian springs. In addition to these native species eighteen species of introduced fish are found in various freshwater habitats throughout Australia.

Until the beginning of European settlement Australian freshwater fish had little to fear except their natural enemies. Since then the grading of riverbeds and courses and the creation of dams, the discharge of effluent from industry, the run-off from chemical fertilisers and pesticides used in agriculture and the introduction of foreign species such as trout and carp have all irreversibly changed many freshwater habitats, playing havoc with fish life-cycles and populations and placing some species on the endangered list, although as yet none are known to have been made extinct. Now that these destabilising influences are recognised it can only be hoped that they will be avoided in the future.

Frogs (order Anura) are the only amphibians — a class shared by two other groups of animals, the worm-like caecilians and the newts and salamanders — found in Australia. Confusion still exists about the proper common names for various species of frogs; warty terrestrial creatures are generally called toads, while smooth-skinned mostly arboreal species are usually referred to as frogs. Strictly speaking, only one true species of toad exists in Australia and this is the cane toad (Bufo marinus) which was introduced to Queensland from Hawaii in 1935 and has since spread into New South Wales. The 180 odd species of frogs found in Australia illustrate an incredible diversity of coloration and form and possess many specialised adaptations to different habitats. Some of these adaptations are demonstrated by the various ways in which frogs keep their skin moist — an essential prerequisite for survival as frogs acquire much of their oxygen directly through the skin and moisture is a necessary part of this process. Many species only emerge from cover at night, thus lessening the chances of dehydration

The massive **Murray cod** (*Maccullochella peeli*) is most likely to be seen in the early morning or late afternoon as it hunts at night and spends most of the day hidden in hollow logs or underneath snags. This species breeds inside or on top of logs and guards the eggs until the larvae hatch.

This **central Queensland watercourse**, bordered by paperbarks, is typical of the region's smaller rivers. It has dried up during the dry season so that the water is limited to permanent pools, connected by small rapids. Despite the scarcity of water the pools provide a habitat for blue-eyes, catfish, gudgeons, hardyheads, long toms and other species of fish. Indigenous reptiles include long-necked and snapper tortoises, children's pythons, water skinks, goannas, such as the lace monitors, carpet and tree snakes and even the deadly taipan.

This flat **mulga country** is dotted with acacia trees that stretch into the distance. Despite the dry conditions and its often desolate appearance, several species of lizards and snakes live here, even burrowing species of frogs will come out of their burrows, should heavy rain inundate the landscape.

This **rainforest** fringes the shore of Lake Barrine in the Atherton Tableland and is characteristic of almost any tropical river frontage in north eastern Queensland. The luxuriant rainforest growth includes umbrella trees, palms, liannas and various hardwoods and serves as a habitat for an abundance of cold-blooded animals. Common representatives include numerous species of frog and snakes. One of the most prominent reptiles is the Amethyst python which can sometimes be found sunning itself on exposed branches or flattened reeds close to the water during the winter months. Fish include grunters, jungle perch, blue-eyes and several species of rainbow, gudgeons and catfish.

due to evaporation, while others spend much of their lives in suspended animation underground, only surfacing to eat and breed during or immediately after rain. Defence requirements are also dealt with differently. While many frogs rely solely upon flight for protection against enemies, others have evolved bright 'danger' colours, such as red, yellow or black or combinations of these hues, to warn off predators and some species make themselves taste unpleasant by secreting poison from skin glands.

Australian frogs are divided into four families. The largest group is the southern frogs (family Myobatrachidae) which includes some 100 terrestrial and burrowing species found throughout the country. Members of this family are sometimes highly specialised with individual species, such as the corroboree frog (Pseudophryne corroboree) and the trilling frog (Neobatrachus centralis), being encountered in habitats as diverse as alpine bogs and arid deserts respectively. Australia's seventy or so species of tree frog (family Hylidae) range from arboreal species equipped with suction pads on their fingers and toes to terrestrial creatures with longer limbs. A recent inclusion in this family is the water holding frog (genus Cyclorana), a burrowing species found in dry habitats. While both the southern and tree frog families are thought to have been present in Australia for some time, thus accounting for their wide distribution and variety of specialisation to different habitats and conditions, the narrow-mouthed and true frog families are regarded as relatively recent arrivals from the north. Only eight species of narrow-mouthed

frogs (family Microhylidae) are known. This family includes both terrestrial and arboreal creatures and is mainly confined to the rain forests of north east Queensland. The exception to this distribution is Spenophryne robusta which also lives in the monsoonal rain forests of Arnhem Land. Narrow-mouthed frogs are rarely seen as they spend most of their time well concealed amongst leaf litter and debris on forest floors. The true frogs (family Ranidae) are regarded as one of the most abundant and widespread frog families in the world. However, they are represented in Australia by a solitary species, the wood frog (Rana daemelii), a terrestrial semi-aquatic creature found along the eastern side of the Cape York Peninsula.

The third class of cold-blooded vertebrates are the reptiles, a category that includes crocodiles (order Crocodilia), tortoises and turtles (order Chelonia), snakes and lizards (order Squamata) and the New Zealand tuatara (Sphenodon punctatus), the sole surviving member of the order Rhynchocephalia. Reptiles evolved from an early form of amphibian and reached their peak, in terms of diversity and pre-eminence, some 120 million years ago with the dinosaurs. Although greatly reduced in number and type since then the reptiles are still the largest group of cold-blooded vertebrates found in Australia and the rest of the world. The largest reptiles present in Australia are the crocodiles (family Crocodylidae), which are represented in this country by two species, the freshwater crocodile (Crocodylus johnstoni) and the saltwater or estuarine crocodile (Crocodylus porosus). Both have thick armoured skin, powerful tails and four-chambered hearts and live in tropical waters in the north and north east. Their individual characteristics and habits are discussed in the Timor Sea region.

Out of the turtles and tortoises only the latter may be properly considered as freshwater creatures. Because of the many similarities between these animals and the uncertainty this has created in regard to their common names, it is necessary to point out their fundamental differences. Briefly, in Australia turtles are said to be large marine creatures with flipper-like legs while tortoises are land and freshwater animals with toed feet which are webbed. In addition turtles draw their heads straight back under their shells when threatened while Australian tortoises retract their heads in a sideways manner. The exception to this rule is the pig-nose turtle (Carettochelys insculpta), a freshwater reptile found in the Timor Sea region that is able to pull its head straight back. Tortoises have several habits in common, particularly with regard to their breeding behav-

iour. After mating with the males in water, the females crawl ashore in spring or summer to lay their eggs in burrows dug above the high water level. Hatching usually takes from several weeks to a few months, depending on environmental conditions, and after hatching the young tortoises quickly head for the sanctuary of the water where they remain for several months, algae growing on their shells.

Despite this camouflage juveniles are preyed upon by numerous enemies but adults face fewer predators barring crocodiles, feral pigs, dingoes and large birds of prey, such as sea eagles. Although tortoises spend most of their time in or near water, some species are able to migrate across land in search of fresh habitats and others are able to survive lengthy droughts by remaining buried in mud.

Broadly speaking the tortoises found in Australia may be divided into two categories, the long-neck or snake-neck tortoises and their short-neck relatives. Both groups belong to the family Chelidae and collectively they number some sixteen species, with six types having long necks and ten possessing short necks. Snake-necks are easily recognisable but short-necks are harder to differentiate from one another, particularly as this group sometimes shows great variation within a single species. While some species merely alter colour to suit different habitats the red-face tortoise (Emydura australis) actually changes colour within its lifetime, so that adults look different from juveniles. Several other distinctions may be drawn between snake-necks and short-necks besides the length of their necks. While the former prefer quiet backwaters, lagoons and billabongs the latter may sometimes be found in swift-flowing streams and rivers. Snake-necks are aggressive carnivores, ambushing or stalking molluscs, crustaceans and fish, while short-necks enjoy an omnivorous diet that includes aquatic vegetation. Finally, while snake-necks have no readily apparent external sexual characteristics, male short-necks are singled out from females by their longer tails.

The largest group of reptiles are the lizards which may be divided into five families: skinks, geckos, dragons, goannas or monitors and pygopods or snake-lizards. The most plentiful of these are the skinks (family Scincidae), with some 300 different species being recognised although many are very similar in appearance and, except to experts, are difficult to tell apart. Varying enormously in size, from the sixty centimetre long land mullet (Egernia

Salt lakes, like this one in Western Australia, opposite, are most common in central and western Australia. Although usually little more than dried out salt pans — Lake Eyre has only been flooded twice since the white settlement of Australia — heavy rain causes them to flood, resulting in an explosion of wildlife, particularly fish. However, once evaporation causes salinity to increase, the fish which provide food for many waterbirds, perish and the birds running short of food move away again.

This **bushfire**, photographed in the Kimberley region, swept through the landscape very quickly, burning the grass but leaving the trees practically unscathed. Bushfires, far from being automatically destructive to trees, often play a crucial role in their regeneration, exploding seed capsules and dispersing their contents. When fires occur during the middle of the day many cold-blooded animals are in their burrows and thus escape the flames.

The **South Alligator River** in the Northern Territory is typical of that region's rich alluvial flood plains. During the wet season when the rivers burst their banks, engulfing the surrounding plains, the area teems with wild life, from crocodiles, to lizards and snakes, to frogs and fish.

The **emerald speckled frog** (*Litoria peronii*), opposite, is a common inhabitant of Australia's south eastern coastal region. It is a nocturnal animal and is most likely to be found sitting quietly in trees hunting insects. During dry spells emerald speckled frogs may sometimes be seen massed around rural out-houses and toilets in search of moisture. The frog's yellow and black groin serves as a defensive mechanism. Predators are sometimes momentarily frightened by a flash of these colours when the frog suddenly leaps away, giving the frog an opportunity to escape.

The **spiny-tailed gecko** (*Diplodactylus ciliaris*) is found throughout much of northern Australia, from the Kimberleys to Arnhem Land. It is often seen crossing roads at night and is less frequently observed crouching motionless on the bark of trees, where its camouflage is most effective. It is named after the soft spines on its tail which can eject a sticky irritating substance, suitable for spraying in the eyes or mouth of predators, such as snakes, small carnivores and birds. Like all geckos, the spiny-tail gecko is nocturnal and it eats small insects and spiders.

major) to numerous creatures a few centimetres in length, skinks may be found in almost any type of habitat and are one of the most frequently seen lizards. Common species include the common garden skink (Cryptoblepharus virgatus) and various blue-tongue lizards (genus Tiliqua). Most skinks possess long cylindrical bodies of varying thickness, movable eyelids, broad blunt tongues and, in many cases, greatly reduced legs. Some skinks have such short legs that they slither along in similar fashion to a worm or a snake. These skinks are usually burrowing types that lead cryptic lives in loose sand or soil. Skinks may be either nocturnal or diurnal and most species hunt insects although some of the larger species, such as the shingle-back or bob-tail (Trachydosaurus rugosus) and the eastern blue-tongue (Tiliqua scincoides), supplement this diet with berries and flowers. While the smaller skinks are usually well camouflaged and rely upon speed and agility to get them out of trouble some of the larger and slower species resort to bluff when faced by predators. For example, blue-tongues attempt to look fiercer than they really are by suddenly opening their mouths when cornered to reveal startlingly blue tongues. Different species of skink reproduce either by laying eggs or else by giving birth to live young.

Geckos (family Gekkonidae) are found in all parts of Australia except Tasmania and occupy both terrestrial and arboreal habitats. Some have a velvety appearance and are covered with loose scaly skin which often looks translucent, but others have quite rough skin. Probably the most commonly encountered species of the seventy-five or so types of gecko is the house gecko (Gehyra australis) which is frequently observed running around walls and ceilings in northern Australia. In recent years though this species has been supplanted in some areas by an introduced species from Asia. Although geckos are generally thought of as being able to walk along vertical surfaces or even upside down with impunity, in reality only a few species are able to do this. Unlike frogs, geckos lack suction discs and instead grip these surfaces by using very fine hooks located on lamellae on their toes. It is fallacious to think that geckos can adhere to perfectly flat surfaces though as the hooks need something to grip. Even glass, which appears flat, is pitted with numerous microscopic indentations. Most geckos use their claws when climbing.

Along with their reputation for agility geckos are also renowned for their habit of jettisoning their tails during stressful situations, particularly when fighting amongst themselves or during confrontations with enemies. This strategy is mainly used as a form of diversion; the discarded and still wriggling tail is intended to momentarily capture the aggressor's attention while the gecko, sans tail, escapes. A replacement tail is grown later. Other geckos attempt to defend themselves by ejecting an irritating substance from glands in their tails. Geckos are carnivorous and stalk their prey — usually composed of insects, although the larger species sometimes eat smaller geckos — with cat-like stealth, nervously twitching their tails as they edge towards their quarry before suddenly pouncing and, if successful, swallowing the victim whole. Geckos are often observed washing their eyes with their tongues, a necessary habit as these lizards lack eyelids. Instead the eye is covered with a protective lens which is cast off at the same time as the gecko sheds its skin. Because Australian geckos are nocturnal they also possess large round eyes, with the pupils shrinking to vertical slits during the daytime to minimise the amount of light let onto the retinas. These lizards reproduce by laying eggs, either depositing hard-shelled eggs under bark or in crevices or else laying parchment-shelled eggs in burrows, carefully blocking the entrance to these holes once the eggs have been laid. Clutch sizes vary with different species and the hatching time depends upon climatic and other environmental factors. Finally, unlike other reptiles, geckos are distinguished by their ability to make sounds such as barks.

Dragons (family Agamidae) belong to the agamid, or Old World, group of lizards and closely resemble iguanas although the two types of reptile are in no way related. About sixty species of dragons exist in Australia and these lizards are characterised by big rounded heads, stout bodies, powerful limbs, long, tapering tails and rough scales. They are also diurnal in habit and while most dragons are terrestrial a few species are arboreal or aquatic. Dragons are found throughout Australia and are most at home in dry desert-type environments, being particularly fond of sitting atop fence posts or termite hills and sunning themselves. These carnivorous creatures sometimes live in communities with several females being gathered around a single male and males often engage in sparring displays during the breeding season. Dragons are egg-layers and females deposit their parchment-shelled eggs in burrows, with hatching times taking at least three months depending on environmental conditions. This family contains some of Australia's most spectacular lizards, including the mountain devil (Moloch horridus), the bearded dragon (Pogona barbatus) and the frilled-neck lizard (Chlamydosaurus kingii). When menaced by enemies most dragons quickly attempt to escape, running away or else disappearing into burrows, but a few species stand their ground and try to look fierce by putting on fantastic displays. The frilled-neck lizard erects a large frill of skin like an umbrella around its neck and the bearded dragon raises spike-encrusted skin which normally lies in loose folds around its throat. Both combine this behaviour with aggressive hissing and open their mouths to reveal rows of sharp teeth that can give painful bites if necessary.

The **banjo frog** (*Limnodynostes terrareginae*) is widely distributed. It grows to about 80 mm in length and is usually encountered after heavy rain at night. It eats most insects and is heavily preyed upon by tiger and black snakes.

The **green tree snake** (*Dendrelaphis punctulatus*), one of Australia's few harmless snakes, is most commonly found near water along the east coast and in the north of the country. Its coloration varies, ranging from blue, through yellow and green to black with a yellow belly, but not apparently, as a response to any particular habitat as different coloured tree snakes co-exist in the same environment. This blue specimen was seen near Tinnaroo Dam in the Atherton Tableland.

The **Townsville blue-eye** (*Pseudomugil signifer*) is a comparatively rare fish and possibly even a new species. Although it can survive in fresh water it has only been found so far in the brackish sections of the Ross River, in tributaries near Townsville and in creeks on Palm Island. Commonly about 5–6 cm long, this blue-eye is possibly the most striking of all blue-eyes. During the breeding season the males engage in sparring displays, during which they race through the water with their fins erect to breaking point.

The **shingle-back lizard** (*Trachydosaurus rugosus*) usually inhabits semi-deserts, or savannah and open forests throughout the southern part of Australia. Although omnivorous, this lizard, which can grow up to 40 cm long, is especially fond of flowers and may sometimes be seen grazing on newly emerged flowers after rain. It is preyed upon by feral cats and foxes and when cornered adopts a U-shaped stance, aggressively flicking its tongue.

Australia's largest lizards belong to the goanna or monitor (family Varanidae) clan, which includes seven species over one metre in length. The biggest goanna of all is the magnificent perentie (Varanus giganteus) which grows to over two metres. However, the other seventeen species are much smaller with the short-tailed goanna (Varanus brevicauda) reaching a mere twenty centimetres. Despite these considerable differences in scale goannas have several physical features in common and are distinguished from other lizards by their loose rough skin, long necks and bodies, powerful limbs and forked tongues, which flick in and out of the mouth when these creatures are alert. Although a few species are arboreal or semi-aquatic in habit, most goannas are terrestrial and they are most likely to be seen on the ground during the early morning or late afternoon when the temperatures are more to their liking. Goannas are carnivorous and, depending on species, eat a wide variety of foods from insects to small mammals and carrion. Many goannas are fond of eggs and to this end rob crocodile, bird and tortoise nests. When molested or threatened these lizards usually run away but if cornered they will stand high on their legs and swing their tails in an attempt to immobilse the enemy. All goannas are egg-layers, depositing their eggs in burrows, termite mounds or hollow trees.

The final group of lizards are the pygopods or snake-lizards (family Pygopodidae). The thirty or so species that comprise this endemic Australian family are generally small and snake or worm-like in appearance. Some pygopods are diurnal but most species are nocturnal and lead cryptic lives. The larger and more commonly encountered species may sometimes be mistaken for snakes — a factor exploited by some pygopods during confrontations with enemies — but closer examination shows that they, along with the other members of the family, are distinguished from snakes by blunt, fleshy tongues, usually visible ear openings, scaly flaps on either side of the vent that are the remnants of hind legs and tails that are easily shed when the pygopod is molested. Pygopods are thought to be closely related to geckos. Most pygopods are rather plain in colour or else have stripes running the length of their body. They are mostly terrestrial creatures and are distributed throughout the drier parts of Australia, feeding on insects, skinks and geckos. Pygopods are egg-layers.

Unlike the non-venomous lizards most of the 130 or so species of snake found in Australia are poisonous, although less than twenty are regarded as dangerous to humans. All are differentiated from other reptiles by their combined lack of legs, eyelids and ear openings and all snakes have forked tongues which are flicked in and out without opening the jaws via a small notch between the lips. The tongue is frequently withdrawn and pushed into two taste pits in the upper jaw known as Jacobson's organs. As snakes are very poor sighted except at close range they rely heavily upon the tongue to relay the whereabouts of food and detect danger. Ground vibrations also warn snakes of danger. Most snakes have a lower jaw that can be disengaged from the skull, allowing the mouth to expand so that they can swallow large pieces of food such as birds or mammals. Teeth are used for gripping not biting and snakes swallow food by moving their flexible jaw steadily forward over the prey. Elastic bodies allow passage of food to the stomach and total

The **chequered rainbow fish** (*Melanotaenia s. inornata*) is one of the most colourful fish found in Australia. It is commonly found in all rivers flowing into the Gulf of Carpentaria from Cape York to Arnhem Land and can survive equally well in flowing and standing water. Breeding normally occurs after heavy rainfalls when flooding causes an abundance of microbic food suitable for the larva. Hoping to attract females, the male rainbows put on a tremendous display, swimming about rapidly with their fins fully erect. Males will dart around the females trying to entice one into vegetation, where they hover side by side, ejecting eggs and sperm. The fertilised eggs are then scattered about and their sticky threads allow them to attach to the surrounds where they hatch after about a week.

digestion may take some time. All snakes are carnivores. Different species reproduce by laying eggs or by giving birth to live young. These reptiles are found all over Australia and occupy numerous sorts of habitat with different species being both nocturnal and diurnal.

Five families of land snakes are found in Australia with the largest group being the family Elapidae, which contains some seventy species of venomous snakes. Most are small, nocturnal and secretive and as such are rarely seen. All are front-fanged and the venom ejected into the prey from the fangs, or hollow teeth connected to poison glands, is usually neurotoxic. Probably the most widespread member of this family is the king brown snake (Pseudechis australis), also known as the mulga snake. This species is related to the red-bellied black snake (Pseudechis porphyriacus), found on the east coast. Other large and dangerous elapids include the various types of brown snakes (genus Pseudonaja), which are fast-moving and nervous, and two species of tiger snakes (genus Notechis), which are found in southern Australia including Tasmania and several islands in Bass Strait. Several of these highly venomous creatures possess black colouring as an adaptation to their temperate habitats and they are often encountered in moist places. Copperheads (Austrelaps superbus) frequently share the same habitats as the tiger snakes. Three species of death adder (genus Acanthophis) are found in Australia. These snakes are so named because of their physical similarity to the European adder or viper but no relationship between the European and Australian snakes exists. Death adders are considered to be one of the most dangerous elapid snakes because of their fairly large fangs, extremely toxic venom and excellent camouflage. All three species use their worm-like tails as a lure to attract prey. Australia's largest and most venomous elaphid snake is the taipan (Oxyuranus scutellatus) which can grow to well over two metres. This species is found in the northern part of Australia while a close cousin, the fierce snake (Oxyuranus microlepidota), frequents the drier inland regions of the continent.

Although the family Colubridae represents the world's largest group of snakes only ten species of this type are found in Australia. They include three non-venomous species; the keelback (Amphiesma mairii), a semi-aquatic snake, and two green tree snakes (genus Dendrelaphis), found in arboreal and terrestrial habitats in northern and eastern Australia. Several other species belonging to this family, such as the brown tree snake (Boiga irregularis) and the aquatic Macleay's water snake (Enhydris polylepis), are venomous rear-fanged creatures. However, rear-fanged snakes are not particularly harmful to humans except in the case of very large specimens because their teeth are located too far back in the mouth. The family Typhlopidae includes about thirty species of blind snake contained in the one genus (Ramphotyphlops). These non-venomous snakes are fairly similar in appearance, with smooth shiny scales, short tails, uniformly thick bodies, curved mouths located well back on the underside of the head and eyes that have been reduced to small dark marks under the scales. Blind snakes are nocturnal and are seldom seen, spending most of their lives burrowing underground in search of termites and ants in generally dry habitats. Two non-venomous, aquatic snakes represent the family Acrochordidae in Australia. Both live in the extreme north.

The final group of snakes belong to the family Boidae and are all non-venomous pythons. This family includes Australia's largest snakes, such as the Amethystine python (Morelia amethystina), which has been reported as exceeding seven metres in length. These bulky and generally slow-moving snakes coil around their victims — generally warm-blooded animals such as birds or small mammals — and asphyxiate them, later swallowing the prey whole. Pythons reproduce by laying eggs, using their coils to guard and incubate the eggs until they hatch. Most pythons are nocturnal but these snakes may sometimes be observed sunning themselves during they day. Besides the Amethystine python other species include the Children's python (Bothrocheilus childreni), the smallest and most common member of this family, and the well-known carpet snake and diamond python (Morelia spilotes), found in a variety of habitats and colour forms throughout much of Australia. The two species of Aspidites pythons, the black-headed python (A. melanocephalus) and the woma (A. ramsayi), are found in drier regions.

The **Low land copperhead** (***Austrelaps superba***) is most commonly encountered in the regions around Mount Gambier in south eastern Australia. Full grown specimens reach 1.2 m in length and are mainly active during daylight, hunting frogs and small mammals, although they are occasionally encountered at night. The heavy bodied copperhead is fairly sluggish and slow moving compared to the brown snake but it is also venomous. When threatened the copperhead erects the front third of its body, adopting an aggressive cobra-like pose. Though a fairly large snake it has few enemies other than man.

1

THE NORTH EAST

A spotted barramundi *(Scleropages leichardti)* takes a cockroach from the surface. These fish are such fast swimmers and have such keen eyesight that they have been known to catch small frogs from waterside plants. The species lives in the upper reaches of Queensland's Fitzroy River system and, at a fully grown length of 90 cm, is one of Australia's largest freshwater fish. Unfortunately the barramundi's size coupled with its prowess as a fighter have made it very popular with anglers. Stocks are now so low that spotted barramundi are being bred in ponds for restocking purposes.

STRETCHING FROM CAPE YORK southwards to the McPherson Range on the Queensland–New South Wales border, this region is demarcated in the east by the Pacific Ocean and in the west by the Eastern Highlands or Great Dividing Range. It contains eastern Australia's two major drainage systems, the Burdekin and Fitzroy rivers. The longer of the pair, the Burdekin, has two main tributaries — the Belyando and the Suttor — while the Fitzroy is fed by the Dawson and MacKenzie rivers. Unlike most coastal rivers, the Burdekin and Fitzroy are long and meandering with gradual gradients and a north-south flow for much of their course. As such, they have more in common with the waterways which drain towards the Gulf of Carpentaria west of the Eastern Highlands. Both rivers are intermittent in flow during dry seasons, especially in their upper reaches, so that they sometimes shrink to little more than a series of lagoons and pools variously connected by shallow streams. Over 130,000 square kilometres are drained by each river, although the Burdekin has a much greater flow because of summer run-off in the river's lower reaches.

The origin of these serpentine watercourses is uncertain. They were either once westward flowing and forced to reverse their direction by the uplifting of the Eastern Highlands, or else drained towards the sea by a more direct route before being thwarted by the appearance of the coastal ranges. Both rivers possess extensive delta systems, containing numerous channels and inlets fringed by thick mangrove swamps. Numerous shorter rivers, such as the Daintree, Barron, Tully, Burnett, Mary and Brisbane, also flow from the hinterland to the sea. They are both younger and shorter than the Burdekin and Fitzroy and usually possess steeper gradients, so that their contents sometimes spill from the inland tablelands and ranges via dramatic gorges and falls, such as the spectacular 300 metres high Wallaman

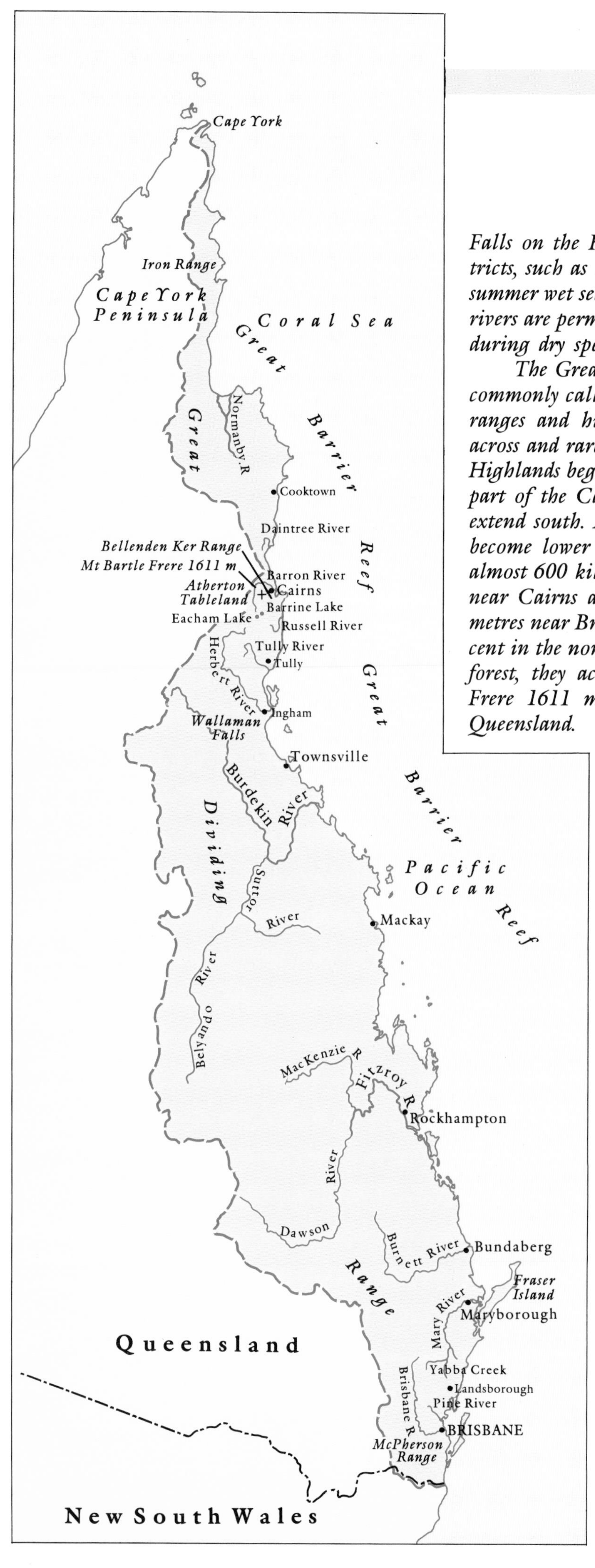

Falls on the Herbert River. Many rivers in high rainfall districts, such as the Atherton Tableland, rise suddenly during the summer wet season and can cause serious flooding. Most of these rivers are permanent although their flow may drop considerably during dry spells.

The Great Dividing Range, as the Eastern Highlands are commonly called, is actually a series of discontinuous plateaux, ranges and high downs sprawling from 150–300 kilometres across and rarely exceeding 1500 metres in height. The Eastern Highlands begin as a narrow spine running through the eastern part of the Cape York Peninsula, gradually widening as they extend south. Around the Tropic of Capricorn the mountains become lower and bulge westwards, so that at times they are almost 600 kilometres from the sea, compared to 30 kilometres near Cairns and, after swinging back to the coast, 120 kilometres near Brisbane. Visually the Highlands are most magnificent in the north and south where, covered with luxuriant rainforest, they achieve their greatest height, with Mount Bartle Frere 1611 metres) near Cairns being the highest peak in Queensland.

Originally formed as a result of folding and faulting of the earth's crust and subsequently much eroded, the Eastern Highlands were also affected by scattered volcanic activity. The volcanoes are extinct today but their active period is recalled by a legacy of crater lakes, volcanic plugs and other physical features, illustrated by such famous landmarks as the Glass House Mountains and Lakes Barrine and Eacham. Between the Highlands and the coast lie several separate ranges, some higher than the Highlands themselves. Some of these ranges, such as the Bellenden, Ker, and McPherson ranges, are notable for their exotic scenery and feature steep cliffs and slopes covered with lush rainforest. Off-shore lies the Great Barrier Reef, a chain of coral reefs, cays and islands almost 2000 kilometres long.

The **lungfish** ***(Neoceratodus forsteri)***, opposite, has remained essentially unchanged in appearance for over 100 million years — well into the period when dinosaurs roamed the earth. The family itself is thought to date back 400 million years, making it the world's oldest living vertebrate, and is unique for having both lungs and gills. However, it cannot survive out of water like some of its African and South American cousins. Unfortunately its present native habitat is threatened by chemical run-off from adjacent agricultural land.

The entire region lies within Australia's tropical and sub-tropical zones, with almost two-thirds located north of the Tropic of Capricorn. Mean annual rainfall generally diminishes from north to south, with the Cape York Peninsula averaging 1778 mm per year, Atherton 1370 mm, Townsville 1090 mm, and Bundaberg 1110 mm, and mostly falls during summer between December and March. Most of the rain falls on the coastal plains and the eastern slopes of the Eastern Highlands or coastal ranges, decreasing in intensity towards the west. The central region around the Tropic of Capricorn receives the lowest rainfall, with Rockhampton averaging 940 mm per year. Because the Eastern Highlands are less mountainous at this point, this area is also influenced by conditions from the drier interior. Temperatures are fairly high throughout the region, decreasing slightly with altitude and lower latitudes, and range from a mean January average of 30 degrees Celsius to 17 degrees Celsius in July. Humidity is greatest in the north during this summer.

The **Mary River**, of which **Yabba Creek** is a tributary, is, along with the Burnett River, the only known native habitat of the Queensland lungfish. This particular section of Yabba Creek contains many wide and shallow places where the water is less than a metre deep — ideal for finding lungfish eggs. The bottom is either sandy or gravelly while the water is reasonably clean. Besides lungfish the creek contains freshwater prawns and shrimps, glass perch, catfish, purple spotted gudgeons, empire gudgeons, rainbow fish, blue-eyes and introduced barramundi.

The region contains several large tracts of rainforest, interrupted by larger expanses of more open and drier savannah and woodland. The pre-European rainforest has been considerably eroded by lumbering and farming but sizeable enclaves still exist between Cooktown and Tully, including the Atherton Tableland; around the Herbert River area; north, west and south of Mackay; between Maryborough and Landsborough; and in the McPherson Range. Smaller strips may be found along watercourses, in sheltered gullies and gorges and on mountain tops in areas of high rainfall. In all these places rainforest rarely penetrates further inland than 160 kilometres from the coast. The rainforest in this region is either tropical or sub-tropical, with the former dominating north of the Tropic of Capricorn and the latter south of the line.

Rainforests, as their name suggests, enjoy higher than average rainfall. Totals vary from 1500–3000 mm per year. Consequently rainforest plants have adapted to this moisture charged atmosphere where the humidity is accentuated by minimal direct sunlight and wind which might encourage evaporation. Except for isolated shafts, very little direct sunlight pierces the forest canopy so that the interior is usually in dark shade. Plants have learnt to germinate and develop in this half-light which ensures the exclusion of parkland species, such as acacias and eucalypts. Because of their lush and humid atmosphere rainforests are rarely threatened by bushfires. This allows these forests to accumulate thick layers of decomposing debris which gradually breaks down to form a rich humus.

Outside the rainforest areas, particularly in the drier central section of this region in the upper basins of the Burdekin and Fitzroy rivers, lie large belts of sclerophyll forest where the dominant trees are eucalypts or acacias. These trees grow apart from one another, rather than forming a continuous canopy, and are generally hardwoods. Their smaller leaves, often twisted away from direct sunlight to reduce

Ceratodus Station, below, has the distinction of being named after the lungfish — *Neoceratodus forsteri*. Conversely, not every fish is famous enough to be immortalised by a railway siding, even if it is in the middle of nowhere! Oddly enough, although the locals know what a 'ceratodus' is, they seem to have never heard of the lungfish.

The **eastern purple spotted gudgeon** *(Mogurnda adspersa)*, left, is found in freshwater from northern New South Wales to Cape York. It prefers to live in flowing water, particularly where the river bottom is rocky, but is also found in lakes on the Atherton Tableland and Fraser Island. The gudgeon is very territorial and will defend its own habitat against other fish of the same species. During the breeding season males clean part of a rock or log for a spawning site and try to attract females. Up to 200 elongated eggs are laid. These stick to the substrate and are guarded by the male till hatching occurs usually after seven days.

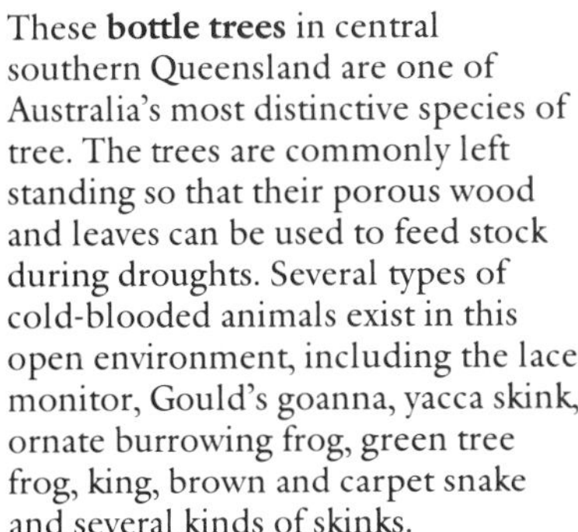

These **bottle trees** in central southern Queensland are one of Australia's most distinctive species of tree. The trees are commonly left standing so that their porous wood and leaves can be used to feed stock during droughts. Several types of cold-blooded animals exist in this open environment, including the lace monitor, Gould's goanna, yacca skink, ornate burrowing frog, green tree frog, king, brown and carpet snake and several kinds of skinks.

The **mouth almighty** *(Glossamia aprion)*, right, overpowers its prey by attacking from ambush. This well camouflaged hunter lurks amongst dense plant growth or underneath logs in still water, waiting for victims which are grabbed and swallowed whole. It is mainly active at night and can eat fish almost equal to its own size. Besides swallowing large fish, the mouth almighty also uses its large mouth as a receptacle in which to store eggs prior to and after hatching.

The **crimson spotted rainbow fish** *(Melanotaenia splendida fluviatilis)*, above, has been one of the world's most popular aquarium fish for almost a century, mainly because of its availability and brilliant coloration.

The **southern soft-ray rainbow fish** *(Rhadinocentrus ornatus)* frequents sandy-bottomed creeks in coastal areas of northern New South Wales and southern Queensland and streams and lakes on Fraser Island. It is generally found in quite large numbers wherever aquatic conditions are constant and feeds on small crustaceans and insect larvae. Like other types of rainbow fish, male soft-rays use flamboyant behaviour, characterised by rapid swimming and colour displays, to attract females during breeding.

transpiration, reflect the adaptations necessary to survive in a more arid environment. Intruding into the parkland and woodland are patches of open savannah grassland which receive even less rainfall. Finally, along the coast, particularly in estuaries and deltas, are extensive areas of mangrove swamp, a generic name given to plants which are able to survive in particularly saline conditions.

Offering a vast array of different habitats, this part of Australia possesses the greates variety of 'cold-blooded' wildlife. Although new species are still being discovered and the taxonomy of some others remains unclear, there are at least fifty-seven species of inland fish and seventy-five species of frogs known to be in the region. Reptiles also inhabit this northern zone. They include: Australian crocodiles, at least nine species of tortoises, 140 species of lizards and over sixty varieties of snakes.

While most 'cold-blooded' animals lead fairly secretive lives, some are well known creatures. Crocodiles, large snakes and highly dangerous venemous snakes make frequent headlines over much of northern Australia. There are however a number of 'famous' species endemic to the north eastern region. One of the best known of these is the Queensland lungfish, a creature from the distant past. Another well known animal is the stomach brooding frog Rheo batrachus silus), which has come to public attention in recent years because of its unique form of breeding. Discovered only ten years ago, it is now feared to be extinct. A different species with the same breeding behaviour was also discovered in early 1984 near Mackay.

Woongoolbver Creek on Fraser Island is a typical rainforest environment. The slightly tea-coloured, fast flowing stream is shallow with a sandy bottom and is home to numerous small fish and crustaceans. It is surrounded by dense rainforest populated by a variety of tropical plants such as native hardwoods, palms and ferns and is inhabited by animals such as skinks, water dragons, snakes and frogs.

The **ornate burrowing frog** *(Limnodynastes ornatus)* is widely distributed throughout eastern inland Australia, yet, like most burrowing frogs, is rarely seen. It is most likely to be spotted during or after rain when it congregates in large numbers to breed in temporary ponds. The breeding cycle, from larva to frog, is very rapid, taking place before the water evaporates.

The **common green tree frog** *(Litoria caerulea)*, is widely distributed throughout Australia, ranging from southern New South Wales to northern Queensland and the Kimberleys. It is adapted to all kinds of habitats and is often seen on roads or around rural buildings at night. At 12 cm in length the common green tree frog is one of Australia's largest frogs. It has a distinctive loud and deep call and although usually bright green in colour, is occasionally brown or even blue. The specific scientific name 'caerulea' meaning 'blue' comes from the original preserved specimen which had lost the yellow pigment from its skin and hence looked blue when it was first described.

The **honey blue-eye** *(Pseudomugil mellis)*, left, is one of Australia's smallest fish, averaging just 25 mm when fully grown. It is found in swampy, dark water creeks along the coast north of Brisbane. Unfortunately the blue-eyes natural habitat is being destroyed by urban and tourist development and the species is now endangered. Blue-eyes are active during daylight when they feed on insect larvae and small crustaceans. The behaviour of the male during breeding is similar to that displayed by various species of rainbow fish and involves rapid swimming and visual displays designed to attract females . Shown are two males sparring along the border of their territories.

Tryon's gecko *(Oedura tryoni)*, below, inhabits rocky outcrops and large trees throughout its range which encompasses the highlands of northern New South Wales and south eastern Queensland. It is a very variable coloured gecko, making its home in large rainforest trees on Fraser Island. A nocturnal lizard reaching 15 cm in length including the tail. Tryon's gecko feeds mainly on arthropods which are stalked and overpowered in cat-like fashion.

The **yakka skink** *(Egernia rugosa)*, right, is found in central Queensland, frequenting open forest country characterised by rocky outcrops. It is one of Australia's largest skinks, reaching 35 cm in length and varying in colour from a grey-brown to a chocolate reddish-brown. Little is known about this particular skink's habits. It appears to be most active during the early morning and late afternoon when it hunts for insects. It is possible that this species, like other skinks, supplements its diet with berries.

The diminutive **Bynoe's gecko** *(Heteronotia binoei)* is found throughout virtually the entire Australian continent except for the south east corner. This particular specimen can be encountered in open woodland amongst litter. Usually around 80 mm in length, these geckos are highly variable in their colouring, depending on their environment, and range from black to dull brown to reddish brown and pale grey. Most specimens are covered with lighter or darker variegated or banded markings.

The **golden-tailed gecko** *(Diplodactylus taenicauda)* is an arboreal or tree-living gecko that lives in native cypress forests in northern New South Wales and southern Queensland. It is particularly well camouflaged for this environment and has rarely been detected elsewhere. During the day the golden-tailed gecko clings motionless to cypress branches with its head pressed down in an effort to escape detection. It hunts at night, searching for nocturnal insects. When attacked it squirts a sticky fluid from glands on its tail.

The **water dragon** *(Physignathus lesueuri)*, right, may look fierce but is, in fact, rather a timid creature. Because of its shyness it is rarely seen — the first indication of its presence is often a sudden plop as it dives into water for safety — although they can become quite tame if living in areas of human habitation. It is usually encountered in vegetation alongside creeks and rivers and is distributed from Victoria to Cape York, east of the Great Dividing Range. They vary in coloration; show here is the form typical in south eastern Queensland.

The **Russell River** is a typical example of one of North Queensland's coastal streams. Originating in the Atherton Tableland, the Russell River is fairly short. Its banks are lined with dense rainforest and grasses, with the latter occasionally growing over and under the water, the bottom being gravelly or sandy in the lower reaches. Besides numerous fish and reptile species this type of habitat also supports a growing number of salt water crocodiles, which are becoming more common now that they are a protected species.

The **sooty grunter** (*Hephaestus fuliginosus*), opposite, one of northern Queensland's best known sporting fish, is found in coastal streams between the Burdekin River and Cape York and across to Arnhem Land. It prefers flowing water and is rarely found in billabongs. Growing up to 40 cm in length, the sooty grunter preys on crustaceans and any fish that it can overpower but is in turn eaten by larger fish, such as silver barramundi (*Lates calcarifer*). Sooty grunters vary greatly in colour from slate grey to various shades of brown and are active during daylight.

The minute **Gertrude's blue-eye** (*Pseudomugil gertrudae*), above, is found in small overgrown creeks and swampy areas between Ingham and Cairns. It is active during daylight and feeds on insect larvae, such as mosquito larvae, and small crustaceans. Because of its size — full-grown blue-eyes reach a mere 25 mm in length — the Gertrude's blue-eye is eaten by most predatory fish.

The **dwarf rainbow fish** (*Melanotaenia maccullochii*) is a favourite freshwater species amongst aquarists all over the world. Growing up to 7 cm long, this fish inhabits clear coastal streams between Ingham and Cooktown. One of the most peculiar characteristics of this species is displayed by the male during the spawning season. In order to attract females, the male develops a bright yellow band between the snout and the dorsal fin. This colour can be 'switched' on and off at will and during the mating procedure the male flutters around potential mates like a butterfly, periodically darting off into darker water to demonstrate his neon-like colouring.

The **northern soft-ray rainbow fish** (*Cairnsichthys rhombosomoides*), below right, is only known from a limited number of locations in Queensland's Russell River and its tributaries. It favours small rainforest creeks with soft, clear, swift-flowing water and feeds on small crustaceans and insect larvae.

The striking looking **Cairn's blue-eye** (*Pseudomugil signifer*) is possibly a yet undescribed species. It is commonly found in fast-flowing rainforest streams in the vicinity of Cairns where it can be seen darting around after small crustaceans and insect larvae. Males are very territorial and when kept in aquariums the dominant male will kill other males. The breeding habits of this species are similar to those of other blue-eyes and rainbow fish and females lay large eggs which stick to plants or other submerged matter. Fully grown Cairn's blue-eyes reach 50 mm in length.

The **snake-headed gudgeon** (*Ophieloeotris aporos*), one of the more colourful species of gudgeons, is distributed along the east and north coasts of Australia. It lives in many types of freshwater environments, ranging from estuaries to rainforest creeks, and is usually found under snags or amongst underwater tree roots. This fish is active during both day and night and mainly preys upon crustaceans, although it will eat anything it can overpower. Its colour varies with habitat and mood.

The birth of the lungfish

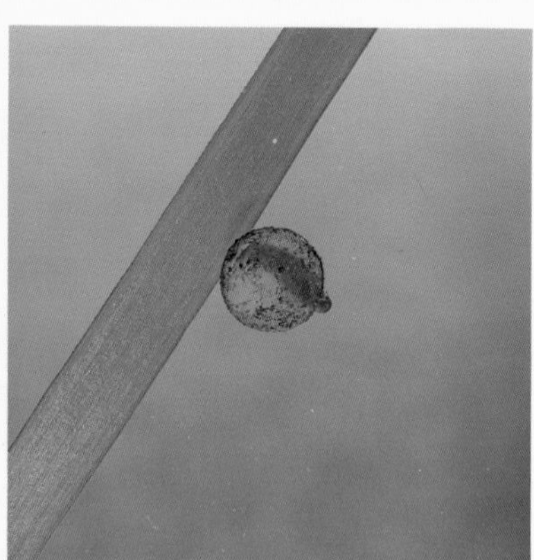

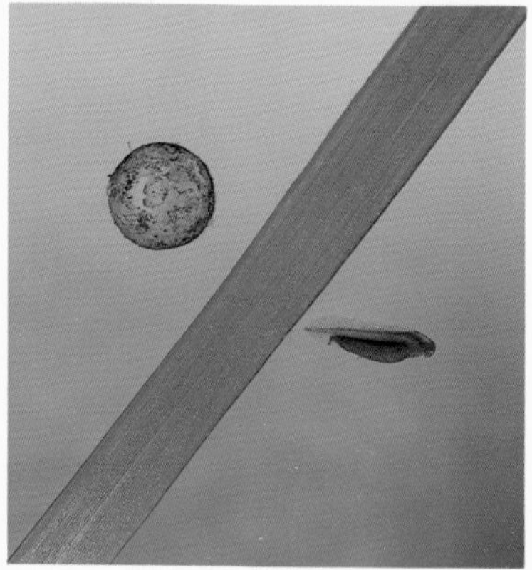

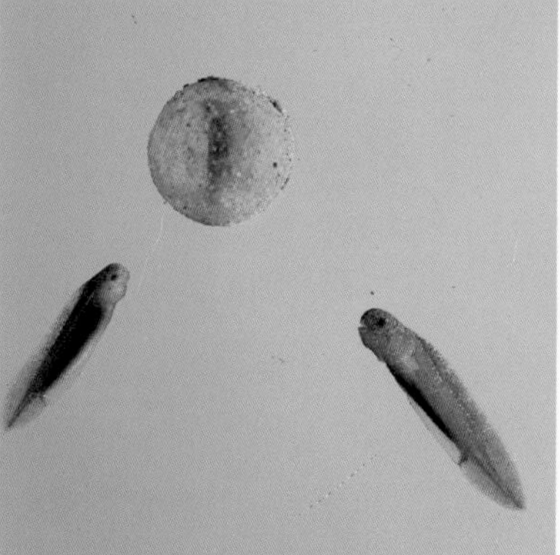

A **Lungfish larva** just emerging from the egg. Very little is known about the breeding habits of lungfish. Breeding occurs between August and December. The male chases and nudges the female amongst waterplants until the eggs are ejected. They are then fertilised by the male. The fertilised eggs, which are about the size of a pea and similar to frog's eggs, are then spread amongst aquatic plants. A sticky, jelly-like substance on the outside of each egg enables it to stick to the plants and not float away. Algae and other floating matter adhere to the egg, providing camouflage. Eggs collected at Yabba Creek and placed under controlled conditions in a fish tank took three weeks to hatch. Once the eggs have hatched the larvae spend most of the time lying on their side, making flicking movements whenever they are disturbed. The photograph on the right shows the lungfish egg before it has hatched and the growth of the juvenile once it has emerged from the shell. The larva on the left is thirty-six hours old while the larva on the right is nine days old.

Lungfish are quite large and may reach up to 1.8 m in length. Although they are occasionally seen during the day they are mainly active at night, when they feed on plant matter and small crustaceans, such as freshwater prawns and shrimps. They also sometimes devour sleeping fish. Their sense of smell is excellent, enabling them to detect their quarry from a considerable distance. Feeding is a protracted process as lungfish suck their food in and out of the mouth, covering it with slime and carefully chewing it, before swallowing. They normally swim slowly along the bottom, with undulating movements of their large tailfin. However, when disturbed, lungfish can move like torpedoes, causing havoc to anything in their path.

The **sawshell tortoises** (*Elsaya latisternum*) in this picture are waiting to be fed with fish offal. Like Krefft's tortoises, these tortoises are fairly common on Queensland's east coast and this species may be found both in lowland rivers and in colder mountain streams. Unlike turtles, which have flippers, tortoises have legs with webbed toes.

Krefft's tortoise (*Eymdura krefftii*), left, is possibly the most common species of tortoise found in the east coast drainage region between the Burnett River and Cape York. It is found in all kinds of habitats and being active during the day it can be observed readily swimming about, or sunning itself on rocks and logs. Freshwater shrimps and prawns, sick fish, carrion and vegetable matter form the Krefft's tortoise's diet. In turn, it is eaten by crocodiles, large fish and by birds of prey while in its juvenile stage. In the past it was also an important food source for Aborigines. Juvenile specimen like this 12 cm long individual shown here spend most of their time in the water and their shell often becomes covered with a thick growth of hairy algae adding to their camouflage. However, as tortoises, like snakes and lizards, shed their skin periodically, this algae growth comes off whenever the shells are renewed.

The **scrub python** (*Morelia amethystina*) is Australia's largest snake and there are records of specimens over 7 m long. The average size is about 3.5–4 m in length. It is a very slender snake however, and therefore seems much smaller than the pythons of Asia or the South American boas. It is found on the eastern part of the Cape York Peninsula. It feeds on mammals and birds. Like all pythons, these snakes asphyxiate their prey, grabbing it by the head and then squeezing the prey with their coils. Once the prey is dead it is swallowed whole by the python.

The **fly-speckled hardyhead** *(Craterocephalus stercusmuscarum)* is one of Australia's most abundant freshwater fishes. Up to 12 cm long, it is found in all sorts of freshwater habitats throughout the northern part of Australia. Hardyheads are most active during the day when they feed on insect larvae, vegetable matter, such as algae and rotting plants, and small invertebrates. Because of their numbers they constitute an important part of the food chain, providing food for most other fish species as well as birds and reptiles.

This species of **carpet snake** (*Morelia spilotes variegata*) might well be a yet undescribed species as it has quite different coloration from the carpet snakes found further south. This beautifully patterned, heavy bodied snake grows to about 1.5 m in length and is found in rainforest habitats near Cairns and on the Atherton Tableland. Like all pythons it is non-venomous, instead killing its prey — usually small mammals — by asphyxiation.

The **jewel skink** (*Carlia jarnoldae*) is one of the many small leaf litter dwelling skinks found in the rainforest of north eastern Queensland. Growing up to 10 cm and fairly common, it feeds on small arthropods that inhabit the forest floor. Easily preyed upon.

The **green python** (*Chondropython viridis*) is found in the Iron Range and in other pockets of monsoonal rainforest in the eastern part of the Cape York Peninsula. Although commonly found in New Guinea it is comparatively rare in Australia. Green pythons average 1.5 m in length when fully grown. Juveniles are either bright yellow or reddish with white markings, changing to brown and finally bright green with white as they grow.

The **pink-tongued skink** (*Tiliqua gerrardi*) inhabits coastal rainforest environments from the Sydney area to Cape York. Although occasionally seen during daylight, this skink is mainly active at night, feeding on small invertebrates as well as slugs and snails. Growing up to 35 cm in length, the pink-tongued skink can be found in a variety of colourforms. The one shown here is from the Atherton Tablelands. It is fairly common. When threatened, the skink stands its ground, facing the attacker, hissing and flicking its bright pink tongue.

The **forest skink** (*Tropidophorus queenslandiae*) lives in rotten logs and amongst decaying foliage on the Atherton Tableland and in the surrounding districts. Although this skink is active throughout both the day and the night it is rarely seen unless deliberately looked for. It hunts small invertebrates. Female forest skinks are live bearers, developing eggs inside their bodies, and give birth to two to three young at a time.

The angle-headed, or **Boyd's forest dragon** (*Gonocephalus boydii*) is restricted to the rainforest districts of tropical northern Queensland. It lives in trees and creepers and is active during daylight, feeding on insects and other smaller invertebrates. This dragon grows to about 50 cm and despite its long legs is rather sluggish, relying on its greenish colouring as camouflage to evade predators. The dorsal spikes are more prominent in the male dragon and are always erect.

Cyrtodactylus loisiadensis is Australia's largest gecko — curiously, despite this it has no common name — and is found both in rainforest environments and in dry, rocky habitats. Like all geckos this fairly common species is nocturnal and feeds chiefly on insects. *C. loisiadensis* grows up to 30 cm in length and is very colourful, with yellow and dark-brown banding.

This close-up of *Cyrtodactylus loisiadnesis* shows the gecko's large lidless eyes. The gecko's eyes are perfectly adapted to its nocturnal habits. During the daylight the pupils narrow into vertical slits, while at night they widen into huge circles enabling excellent night vision. Because the gecko has no eye lids, its eyes are protected by transparent scale.

Rainforests and swamps between the Atherton Tableland and Cape York are the home of the **white lip tree frog** (*Litoria infrafrenata*). Its coloration varies, ranging from bright green with white lips and pinkish-white leg bands to brown or blue. This species is distinguished from other green tree frogs by its particularly long hind legs which are best observed when the frog is jumping. It is fairly common and may sometimes be observed crossing roads at night in great numbers, especially during the breeding season. This relatively slender nocturnal hunter is Australia's largest tree frog.

The **leaf-tailed gecko** (*Phyllurus cornutus*), found in rainforest locations between Coff's Harbour and the Atherton Tableland, is Australia's second largest gecko. Named after its distinctively shaped tail and varying slightly in shape and appearance in different habitats, it lives in rocky outcrops and moss-covered tree trunks. Although fairly common, the leaf-tailed gecko is rarely seen because of its excellent camouflage. It feeds mostly on small insects and spiders. Adult specimen with original tails are rarely found indicating that their enemies are many. The tail of the specimen shown here is regenerated.

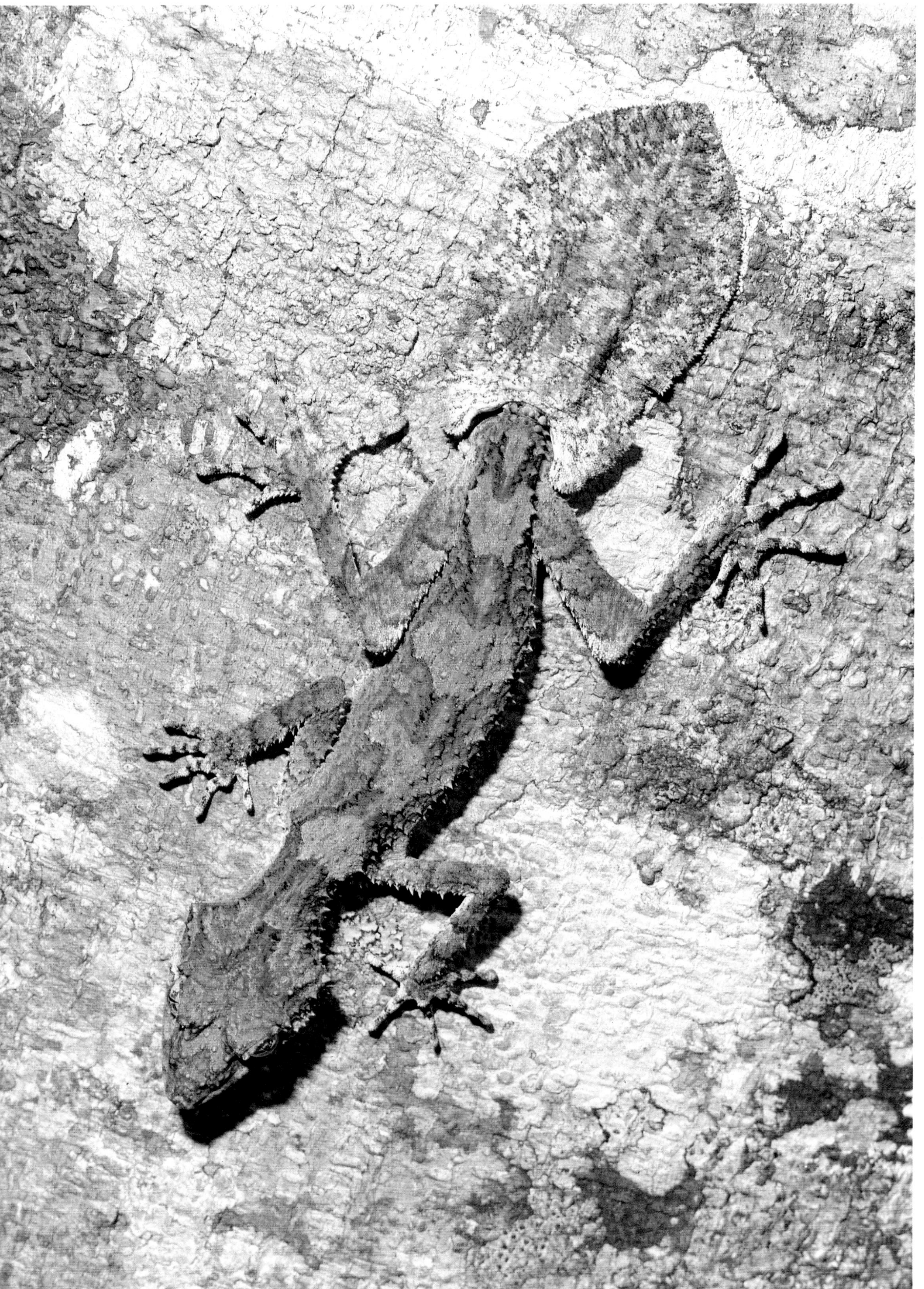

Nyctimystes tympanocryptis is a true tree frog and is rarely found on the ground. It inhabits rainforest areas in the Atherton Tableland region and is a nocturnal hunter, searching for small bugs and spiders. This slender amphibian grows up to 5 cm long and is equipped with very effective camouflage, including a striped lower eye lid that can be drawn over the eye when the frog is sleeping.

At a mere 3 cm in length when fully grown, *Litoria bicolor*, opposite, is one of Australia's smallest frogs. It is distributed throughout northern coastal Australia and frequents river banks, swamps and rainforest habitats. The frog usually has a dull to bright green back and white stomach and an indelible stripe running through its eye and along its side. It is active at night, when it hunts for small insects.

Leisure's tree frog (*Litoria lesuerii*), commonly found along the east coast, is a ground-adapted amphibian, spending most of its time hopping about on the ground. It is most commonly seen on warm, humid nights, or during rain. It grows to a length of 7.5 cm. The black-yellow coloration along the frog's side and on its groin flash whenever it moves, acting as defensive colours.

The **northern river frog** (*Mixophyes schevillii*) is mainly seen beside swift-flowing rainforest streams in the Atherton Tableland region. It hunts at night, eating small invertebrates and providing food for snakes and other nocturnal predators. The frog grows to 9 cm in length.

When threatened the **mangrove monitor** (*Varanus indicus*) reacts like other goannas, employing a mixture of bluff and aggression. After rising up on its legs the goanna tries to exaggerate its size by puffing up its throat and chest, finally thrashing its tail about like a whip. It will also bite and scratch. Mangrove monitors grow up to 1.5 m long and have been known to live for up to twenty years.

The **mangrove monitor** (*Varanus indicus*) inhabits estuaries and coastal forests on the eastern side of the Cape York Peninsula and also occurs less commonly in Arnhem Land. Essentially a water monitor, it has a strongly compressed tail and spends much of its time in the water hunting for food. It will feed on anything from small invertebrates, crabs, fish, small mammals to birds and crocodile eggs. Adult monitors have few enemies other than crocodiles and large birds of prey. The mangrove monitor is active during daylight, mainly in the morning and late afternoon when temperatures are lower.

Mangroves such as these, below, with their characteristic tangle of roots, are commonly seen around much of the northern Australian coastline. They provide habitats for many animals, including the mangrove monitor. 'Mangrove' is a collective name for several species of tree able to survive in the saline conditions of estuaries and coastlines, with different species being suited to different habitats.

The most obvious physical characteristic of **Berney's catfish** (*Arius berneyi*) is its unusually high dorsal fin. It is a schooling fish and as such is normally found feeding at night with other members of the same species on small crustaceans and invertebrates. In turn, it is eaten by larger fish, birds and possibly crocodiles. Berney's catfish is the smallest of the fork-tailed type of catfish and is found in rivers that drain into the southern part of the Gulf of Carpentaria, usually frequenting slightly saline waters just upstream from the river estuaries.

2
THE GULF OF CARPENTARIA

THIS TROPICAL REGION, WHICH occupies a rough crescent around the Gulf of Carpentaria, encompasses the catchment areas of all the rivers which drain into the Gulf itself, stretching from the Roper River in the Northern Territory to the Jardine River on Queensland's Cape York Peninsula. The landscape is largely dominated by the Carpentaria Basin, a large expanse of low-lying terrain dissected by numerous meandering and intermittent rivers. This basin was once much larger and after the last Ice Age extensive areas were inundated with water and turned into the present Gulf of Carpentaria. The basin is broadest at the foot of the Gulf where it sags southwards, following the Flinders River to the Selwyn Range. Except around its periphery, where it is bordered in the east by the Eastern Highlands, to the west by Arnhem Land and the Barkly Tableland and to the southwest by the Mueller Plateau, the Carpentaria Basin is never more than 150 metres above sea level.

Numerous rivers, with a total discharge more than six times that of the Murray-Darling system, flow across this region into the sea. Ranging from east to west the major ones include the Wenlock, Archer, Staaten, Mitchell, Gilbert, Norman, Flinders, Leichhardt, Gregory, Nicholson and Roper. Some, such as the Mitchell and Gregory, rise in high country at the fringe of the region and have a greater discharge than the others. However, when the Mitchell and Gregory reach the plains they become as sluggish and meandering as their companions. Because they generally flow across flat plains with slight gradients many rivers fragment, branching into new channels that frequently intermingle and sometimes even find their way into adjacent rivers. Once the main channels of the Mitchell and Gilbert have splintered into different courses they never coalesce, either reaching the sea separately or else joining forces with the nearby Staaten River.

The longest river in the region is the Flinders, which begins its 840 kilometre trip by

flowing in a north westerly direction before turning north to the sea and meeting the Cloncurry River, its major tributary. Although it flows through a potentially large catchment area the Flinders actually discharges less water than the shorter Mitchell River. This is because for most of its course the Flinders is imprisoned by a narrow flood plain, located lower than the surrounding landscape. As a result the river's channels have become incised and are denied much of the potential run-off which the river would have garnered had it been allowed to wander all over the countryside. The watercourses in the east and west of the region tend to be shorter than those nearer the centre. Many rise in more rugged and broken terrain and commence their journeys through narrow valleys that prohibit extensive sideways exploration until the rivers arrive on the coastal plain. Most rivers enter the sea through estuaries or deltas, sometimes crossing extensive mudflats at their mouths and generally being subject to tidal influences in their lower reaches.

The seasonal rainfall characteristic of this region ensures that many of the rivers are intermittent in flow. Only a few watercourses, such as the Gregory which is fed by permanent springs on the Barkly Tableland, are perennial and even these contain very little water during the dry season. Instead many rivers dwindle into little more than isolated pools or waterholes during the dry season. Often the only reliable sources of water are the region's permanent lagoons and these act as sanctuaries for much of the environment's wildlife during droughts. The onslaught of the wet season dramatically alters the landscape, causing rivers to swell, flood and finally overflow their banks, inundating the surrounding countryside. Sometimes rivers blend together so that parts of the region are covered with vast sheets of water. However, far from being disastrous, this annual advance and retreat of flood waters is an integral part of the landscape's natural rhythm to which the local fauna and flora have both adapted.

The wet and dry seasons are probably the paramount annual events in this region and as such affect everything that lives there. Each is dramatically contrasted to the other and together they take up virtually the entire year so that there are no transitional seasons. The dry season lasts from April to October and during this time dry winds blow from the southeast across the interior of the continent. Very little or no rain falls during this period so that, as the season continues, the landscape becomes progressively more arid and desolate. Rivers shrink, temperatures climb and heatwaves become commonplace. At first the weather is relatively cool but after August both the humidity and the temperatures steadily increase. By October the humidity is almost unbearable so that sporadic but increasingly frequent thunderstorms provide welcome, if short-term, relief. As storms become heavier and more regular from November onwards with the advent of the north west monsoon which blows moist winds across the region, flash floods occur and the landscape gradually becomes sodden and covered with water. By February the wet season has reached its climax and thereafter the rains begin to taper off. Annual rainfall totals for the region vary between 400 mm and 1600 mm each year. They decrease in frequency, reliability and amount towards the interior but increase in the Cape York Peninsula and in Arnhem Land. Hot and humid summers, averaging 35 degrees Celsius in January, and mild dry winters, with a July mean of 30 degrees Celsius, are the norm. Temperature extremes are greater inland.

Generally vegetation becomes less dense and lush with distance from the coast. Because most rain falls during a few summer months the growing season is necessarily short and plants experience rapid growth between November and March. During this period grasses that hardly constitute a stumble in the

Many of the rivers in this region are intermittent so that during the dry season their upper reaches and tributaries shrink to become little more than a series of billabongs or lagoons. The latter are often covered by a carpet of exotically coloured water-lilies and aquatic plants which provide a habitat for numerous microscopic creatures. These in turn form the basis of a food chain ranging from minute fish to giant crocodiles.

Mary Creek, a tributary of the upper Mitchell River, is typical of many Cape York streams which flow westward across the peninsula into the Gulf of Carpentaria. It has a sandy bottom and clear water and alternates in character from a placid, fairly shallow stream during the dry season to a raging torrent during the wet.

The **coal grunter** (*Hephaestus carbo*) is found, usually in schools of ten to fifteen fish, in westward flowing rivers on Cape York and in Arnhem Land. One of the prettiest of all Australian grunters, it likes clear flowing water and is generally encountered amongst snags and submerged logs.

dry season shoot up to heights of over three metres. The main types of vegetation found are sclerophyll (plants with small leaves) forest, sclerophyll parkland (plant cover where the trees are too far apart to form the continuous canopy associated with forests), savannah, grassland, mangrove swamp and monsoonal rain forest.

Mangrove forests, with trees up to ten metres high, are found along the Gulf shores between the Roper and Gregory River mouths and on the southern shores of the Cape York Peninsula, usually where there are estuarine conditions. 'Mangrove' is a common name given to more than twenty species of tree or shrub found in this region. Although separate species of mangrove have evolved different characteristics they all possess certain similarities. All species are remarkably tolerant of salty conditions and have developed highly efficient ways of minimising their salt intake. Some species discharge salt through their leaves while others restrict its entry through their roots. All have found ingenious ways to ingest the maximum amount of oxygen. Some obtain air through small cracks in their aerial roots while others acquire it via root-like extensions. Mangroves are also able to survive in tidal conditions that would destroy other plants. They have even managed to utilise the daily tidal fluctuations to assist their own survival, relying on the tide to disperse seedlings and start new colonies. The tangle of roots associated with these dense forests is used to anchor the trees in their precarious environment, with different species putting out aerial or horizontal roots. The landward side of these forests is sometimes occupied by desolate salt marshes where few plants survive.

Monsoonal rain forest can be found bordering the mangroves in the western part of this region towards Arnhem Land. Because of the seasonal nature of the region's rainfall, which guarantees moisture and humidity for only part of the year, monsoonal rain forest has developed differently from its tropical counterpart. Certain similarities are apparent. Both share many species of tree and each develops a canopy and a layered forest interior. However, monsoonal rain forest canopies are relatively low, rarely exceeding twenty-four metres, and the interiors are much less lush and tangled. This is partly because they lack much of the luxuriant plant life, such as epiphytes, ferns, orchids, mosses and lichens, found growing in tropical rain forests. Monsoonal rain forests

Over six different colour variations of this **banded rainbow fish** (*Melanotaenia trifasciata*) are known in northern Australia and this is the most vivid of them all. This particular form has only been found in rivers that flow into the Gulf of Carpentaria from the western side of Cape York Peninsula and a slight variation in colour is apparent even in this region. Another more slender version of the banded rainbow is found on the eastern side of the peninsula. These fish favour clear flowing streams where the water is soft and acidic. This picture depicts two 12 cm long males sparring during the breeding season.

Until recently **Lorentz's grunter** (*Pingalla lorentzi*) was unknown in Australia and so far specimens have only been collected in a few localities near the western tip of Cape York, although these grunters are found in New Guinea. Like many other grunters, this species swims in clear flowing water and is most likely to be spotted near submerged logs and debris where it feeds on freshwater shrimps and prawns during daylight. This fish grows up to 20 cm in length.

The **yellowfin catfish** (*Neosilurus glencoensis*) is regularly found in freshwater habitats right across the northern part of Australia. It is especially common after the recession of flood waters when thousands of specimens are found in shallow water, providing a feast for hawks, kites and other avian predators. Its poisonous spines, located in the fish's dorsal and pectoral fins, are fairly useless against winged enemies but can deter other fish. Although spasmodically active during daylight, this catfish is most likely to be encountered at night when it searches for worms and crustaceans amongst the detritus along river bottoms. Its coloration varies according to habitat.

The **Jardine River rainbow fish** (*Melanotaenia maccullochi*) has only been detected in small ponds and lagoons along tributaries of Queensland's Jardine River, located at the western tip of Cape York, so far. However, an identically coloured species of rainbow fish is also found in New Guinea's Fly River system. Although similar to another form of *Melanotaenia maccullochi* found in creeks near Cairns, the Jardine River form is smaller, reaching only 4 cm in length, and also exhibits different behaviour during spawning. Small crustaceans and insect larvae provide this rainbow fish with its food and it is part of the diet of most larger fish in this habitat.

also contain hardwood-like eucalypts and numbers of deciduous trees, neither of which appear in tropical rain forest.

Elsewhere the immediate hinterland is covered by dense sclerophyll forests, with eucalypts being the dominant tree species. Forest floors are covered with grass and scrub rather than the carpet of decomposing leaves and matter found in broad-leafed rain forest. The dense coastal forests gradually thin out, giving way first to parkland, then savannah and eventually to grassland. As the trees thin out termite ant hills up to four metres high become an increasingly common sight. Savannah — grassland with scattered trees such as eucalypts and acacias which are able to survive dry conditions — is most predominant in the centre of the region where the Carpentaria Basin runs south to the low Selwyn Range. This frequently flat and monotonous country is covered with clumps of perennial Flinders, Mitchell and blue grass, interspersed with ephemeral grasses and dotted with occasional trees. During summer the grasses die back to their roots or succumb to bushfires before revitalising and shooting up dramatically during the wet season, sometimes reaching three or four metres in height.

It is difficult to determine exactly how many animal species live in this region. Recent investigation into remote areas, like the northern part of the Cape York Peninsula, has revealed the existence of several new fish species, such as the feather-fin rainbow fish (Iriatherina werneri), previously only known in southern New Guinea but now found in the Jardine River. At last count the region contained approximately 125 lizard, 50 snake, 45 fish, 35 frog, 7 tortoise and 2 crocodile species but some of these figures will almost certainly increase as research continues.

The **freshwater sole** (*Brachirus salinarum*) has one extraordinary characteristic — it is able to quickly change its colour, chameleon-like, to blend in with different environments. During this process, the fish's melanin cells, which contain dark pigment, either coalesce or disperse, enabling the fish to turn a dark colour when inhabiting a dark environment or a light tone to suit a light habitat. As might be expected with such a well camouflaged fish, the freshwater sole is difficult to detect, as shown in this picture.

The freshwater sole is rarely found in open situations and frequently burrows into sand beneath rotting leaves. It is found in the upper reaches of rivers draining into the Gulf of Carpentaria and is usually about 15 cm long. This species is nocturnal and feeds on worms and small vertebrates.

Also called the giant perch, the **silver barramundi** (*Lates calcarifer*) is distributed throughout the northern part of Australia, from the Cape York Peninsula across to the Kimberleys, and is found in several habitats, from estuaries to freshwater rivers. In the past the barramundi's reputation as one of Australia's most famous sporting fish caused its numbers to be sorely depleted by both amateur and commercial fishermen. This situation has been partly rectified by recent legislation in some states which protects the barramundi during the summer spawning season. Whereas barramundi measuring up to 1.8 m in length and 60 k in weight have been caught in the past most specimens taken today are about 70–80 cm long. One of the most curious facts about the barramundi is its ability to change sex. Juveniles start life as males but once they reach 50 cm in length, or thereabouts, they turn into females. This juvenile is stalking a rainbow fish.

When the juvenile barramundi arrives at the optimum position for attack, it lunges forward, actually sucking the prey into its huge protruding mouth. Despite its desperate struggles this rainbow fish has no chance of escape. Barramundi are active hunters during both the day and night and will eat anything that they can overpower, from other fish to frogs. Because of their size adults have few enemies, except crocodiles. Barramundi are at their most vulnerable during the juvenile stage.

The juvenile has almost swallowed its catch so that only the victim's tail is visible. The barramundi will now rest under cover until it has digested its meal. The fish's coloration depends on environment. Barramundi that live in saline habitats are silvery in colour while specimens that favour fresh water are yellowy. All juveniles have a white band on their forehead which gradually disappears with age.

This lagoon on the **Mitchell River** was formed as a result of the river shrinking during the dry season. Similar lagoons exist elsewhere on the Mitchell and in the upper reaches of other serpentine watercourses and they eventually dry out altogether when the annual rains are late. However, whilst they retain water, these freshwater lagoons provide homes for a wide variety of wildlife, including tortoises, freshwater crocodiles and numerous types of fish, such as glass perch, rainbows, hardyheads, archer fish and freshwater herrings.

The **northern death adder** (*Acanthophis praelongus*) is one of the most venomous species of snake found in Australia. This specimen has adopted a defensive position, flattening its head and body as it prepares to strike. Although this snake is shown in open country where its vivid coloration makes it conspicuous, the death adder is particularly well camouflaged and usually hides beneath leaf litter instead of crawling away when anything approaches. This habit only makes the snake more dangerous to unwary walkers. When actually touched this snake will strike with lightning speed. Death adders are nocturnal predators and use their thin worm-like tail as a lure to entice birds or lizards. Victims are immobilised within seconds of being bitten and are then swallowed whole, usually head first. This species of snake grows to 45 cm in length.

The **orange nape snake** (*Furina ornata*) is found throughout northern Australia, often in association with termite hills. It is a nocturnal hunter, preferring to spend the daylight hours secreted under debris, and feeds mainly on small skinks. Although venomous the orange nape snake is not dangerous to humans. When threatened it rises up, lifting its head from the ground and striking with its mouth closed. It grows to a length of 70 cm and reproduces by laying eggs.

The **toothless catfish** (*Anodontiglanis dahli*) deals with this apparent feeding disability by sucking food from the river bottom with its mouth, vacuum-cleaner style. As the catfish does have some teeth, its name is really a misnomer; however, the teeth are not situated in the front of the mouth. Using this trawling technique the catfish digs into the bottom right up to its eyes in search of microbes, crustaceans, worms and other creatures. It is most active at night.

Although known to reach 70 cm in length, most specimens rarely grow longer than 30–40 cm and can be found in varying types of freshwater conditions in rivers draining into the Gulf of Carpentaria and in the Kimberley Region. The flesh of this species is highly prized by fishermen.

Like most members of the black snake family, **Collett's snake** (*Pseudechis colletti*) is fairly heavy bodied. One of Australia's most attractively coloured venomous snakes with a striking black and orange coloration it is most likely to be seen in the black soil country around the Gulf of Carpentaria. It is not often encountered, because it is most active at night when it hunts small mammals, such as rats and mice. This 1.5 m long snake has few enemies, other than birds of prey, but when menaced will flatten its body, raise and adopt the striking position.

Although normally very docile, **Spencer's goanna** (*Varanus spenceri*) is quite capable of defending itself if attacked. When threatened the goanna rises high on its legs and puffs out its throat, simultaneously arching its tail to be ready to lash its enemy. Spencer's goannas are most active during the early morning or late afternoon, when they hunt rats and mice or forage for ground-nesting birds' eggs. They are most likely to be observed in the breeding season between August and October. Due to the lack of ground cover in its habitat this species is a keen burrower and the average specimen grows to about 1 m in length.

This flat and rather desolate looking terrain is typical of the **black soil country** found around Julia Creek and Richmond in Queensland's Gulf Country. During the dry season between April and October the land gradually becomes parched as watercourses dwindle and become little more than strings of pools or lagoons. However, the heavy monsoonal rains that fall between November and March transform this environment drastically, causing numerous plants, such as the fluffy pink-topped mulla mulla shown in this picture, to appear. This environment is the typical habitat of Spencer's goanna.

This **centralian blue-tongued lizard** (*Tiliqua multifasciata*) has assumed an aggressive defensive posture, inflating and arching its body, folding its legs and opening its mouth wide, simultaneously hissing and flicking its startlingly blue tongue. If this fails to work the blue-tongue is prepared to go on the offensive, biting its enemy whenever the opportunity presents itself. Usually no longer than 35 cm, the centralian blue-tongue is widely distributed over northern Australia's drier regions, from Queensland to Western Australia, and is fairly common. It dwells on the ground in arid or semi-arid landscapes of the type most commonly associated with spinifex country or stoney and sandy deserts. Coloration varies, ranging from pale grey to grey-brown above and cream or white underneath, with wide orange-brown cross markings from the neck to the tail. This lizard is most likely to be seen in the daytime but may also be active on very hot nights. It once formed an important part of the diet of nomadic Aboriginal people.

Along with other species of burrowing frog, the **northern burrowing frog** (*Cyclorana novaehollandiae*), opposite, survives dry spells by going under ground. To prevent itself dehydrating whilst buried the frog stores water in its bladder and forms a cocoon around its entire body. So successful is this frog at retaining water that it is used by nomadic Aboriginal people as a water supply. Although rarely seen in dry spells the northern burrowing frog can be readily encountered at night during rain, when it emerges from its hole to feed on insects. When menaced the frog inflates itself to appear more formidable. It grows to around 9 cm in length.

One of the smallest and most common pythons in Australia is the **children's python** (*Bothrochilus childreni*), which grows to an average length of 1.3 m. This species was once a member of the *Liasis* genus, a classification which no longer exists. It is found throughout northern Australia in terrain ranging from deserts to forests. Although occasionally seen on trees, this snake is essentially a ground dweller, feeding on small mammals. Children's pythons are extremely variable in colour, although brown is usually the common colour denominator, and recent research has suggested that other similar species may have been erroneously classified as children's pythons. This snake is also distinguished by an iridescent sheen to its skin. If captured the children's python ejects faeces and foul smelling liquid as part of its defensive behaviour.

Widely distributed throughout northern Australia, the **desert tree frog** (*Litoria rubella*) is usually found hidden in deep rock crevices in a number of dry habitats, from scrubland to deserts. This specimen comes from the black soil country around Julia Creek, east of Cloncurry and at 4 cm in length, is one of Australia's smallest tree frogs. Insects form the bulk of its diet.

This landscape is fairly typical of the more arid territory found in the southern parts of the Gulf of Carpentaria region, near the Selwyn Range. Termite mounds, scattered clumps of perennial grasses, such as spinifex and Mitchell grass, and acacia or eucalyptus trees, like the ghost gum in the foreground, stretch away across stoney red soil towards the rocky range in the distance. Several reptiles favour this environment, including the centralian blue-tongued lizard, spiny-tailed goanna, Gould's goanna and king brown snake.

Stoney hills and spinifex dominated sandy deserts are two of the habitats frequented by the **fat-tailed gecko** (*Diplodactylus conspicillatus*), a common inhabitant of the arid inland districts of the Gulf of Carpentaria and elsewhere. Often seen warming their bellies on bitumen roads just after dark, these lizards feed on spiders and small insects, being particularly partial to termites. In turn they form an important part of many larger animals' diets. Unfortunately for the gecko its defensive behaviour is rather rudimentary, relying entirely on deception to confuse the enemy. When attacked the gecko offers the predator its head-shaped tail. If this ruse succeeds the gecko jettisons its tail and escapes.

Most commonly found in the drier inland districts, **Hosmer's skink** (*Egernia hosmeri*) normally lives in rocky outcrops, crevices and scree-covered slopes. The spines that run along its back and sides enable it to wedge itself in cracks and resist the efforts of pawing enemies. Spotted near Cloncurry, it is about 30 cm long when fully grown and is active during the early morning and late afternoon when it hunts insects.

The **sand-goanna** (*Varanus gouldii*), left, grows to about 1.6 m in length and is one of the largest types of goanna found in Australia. The most widely distributed member of this genus, its habitats include desert, mangroves, coastal sclerophyll forest and savannah throughout most of northern Australia. This lizard possesses exceptionally keen eyesight and can sometimes be observed standing on its hind legs and surveying the surrounding landscape in search of foods. This position is also adopted as a defensive stance.

One of Australia's most widely distributed species of goanna is the spiny-tail goanna or **ridge-tail monitor** (*Varanus acanthurus brachyurus*). It ranges throughout northern Australia and is particularly fond of rocky areas which afford protection in crevices along with a regular food supply in the form of other rock-dwelling lizards, spiders and insects. Spiny-tail goannas may also be found living beneath empty bitumen drums along the Barkly Highway west of Mount Isa. This lizard's solitary lifestyle, spikier tail scalations and adult length of 70 cm distinguish it from the other two species of spiny-tail goanna, *V. storri* and *V. primordius*. These lizards generally live in colonies and grow up to 40 cm long. The spiny-tailed goanna is an alert and agile creature and is most likely to be seen in the early morning or late afternoon. It is the most common representative of three sub-species, the other two being *V.a. acanthurus* and *V.a. insulanicus*.

This view shows Arnhem Land's **Jim Jim Creek** and **Falls** immediately before the start of the summer wet season. As with most other watercourses that dissect this rugged plateau, Jim Jim Creek has ceased to flow properly and the waterfall has stopped altogether. Water levels are so low that the large boulders which normally lie under metres of water at the bottom of the creek have emerged. However, there is still sufficient water in this permanent pool to support a lush riverbank vegetation, mostly composed of pandanus palms and paperbarks With the advent of the wet in December this habitat is dramatically altered; the creek turns into a raging torrent and the waterfall plunges down the escarpment face at the head of the valley. Many animals are found in this environment, including freshwater crocodiles, water goannas, several species of snakes and fish such as rainbows, grunters, silver barramundi and archer fish.

3
THE TIMOR SEA

FLANKED BY THE WARM TROPical waters of the Timor and Arafura Seas to the north and by the arid sands of the Great Sandy and Tanami Deserts to the south, Australia's most northerly drainage region incorporates the Top End of the Northern Territory and the north eastern corner of Western Australia. The landscape is dominated by the rugged plateau country of Arnhem Land in the east and the Kimberley Region in the west and drained by the Daly, Victoria, Ord, Fitzroy, Durack and other rivers.

Most of the rivers in this region originate in broken plateau country. To the east in Arnhem Land numerous streams radiate from the central plateau area, flowing south into the Roper River, north into the Arafura Sea and west to link up with the Daly River. Further west in the Kimberleys the watercourses follow a similar pattern, rising deep in the plateau interior and then variously flowing east to join the Ord River, north to the Timor Sea and south west to swell the Fitzroy River. The exception to this rule is the Victoria River which, along with its tributaries, drains the centre of this region. Its headwaters lie in the rugged sand-hill country that borders the desert interior to the south.

Like all rivers that come under the seasonal influence of the north west monsoon, the ones in this region display markedly different characteristics in the wet and dry seasons. During the wet, parts of the region are almost submerged. Numerous streams roar down the steep plateau gradients, carving gullies and gorges and plunging over cascades and waterfalls, before they spill over the plateau escarpments to reach the surrounding plains. When the rivers reach flat ground they fan out, sometimes merging with one another to form vast lakes. Simultaneously the flood waters wash massive quantities of silt down from the higher country, dumping it in the lower reaches of each river and thus steadily extending the region's broad alluvial plains and frequently labyrinth estuarine systems. Most of the rivers

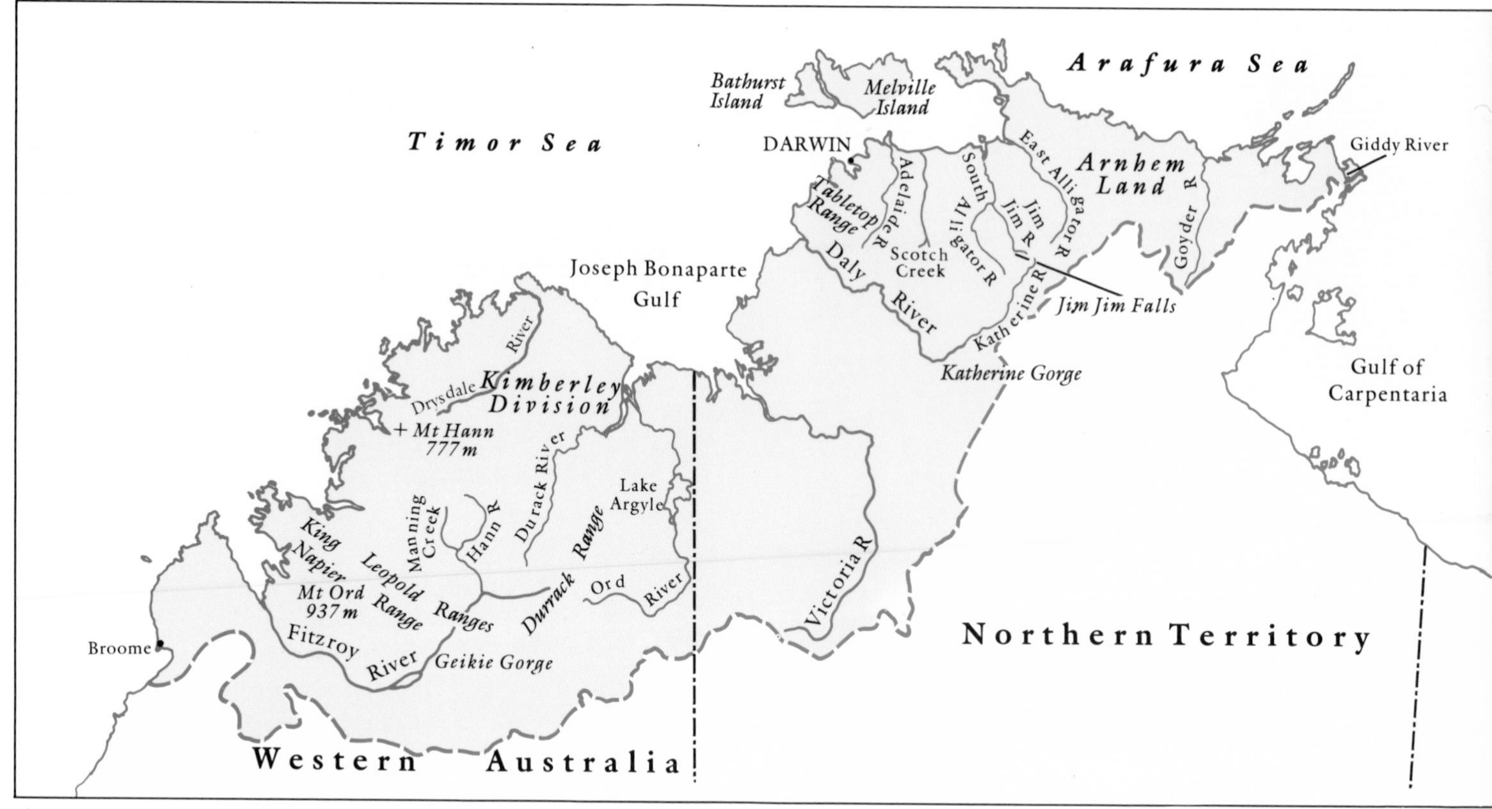

are deep, sluggish and tidal in their lower regions and are often navigable for some distance inland. During the dry season virtually all of the region's rivers undergo a metamorphosis, steadily shrinking in size where they once overwhelmed tree tops until their upper reaches are composed of little more than isolated rock pools and waterholes. The once inundated plains dry out, leaving scattered billabongs and lagoons in place of the sheets of water that lay there during the wet season. The only rivers that manage to escape being intermittent above the furthest limit of tidal influence are those which have their origins in permanent springs.

Throughout much of the region the rivers have had a dramatic impact on the landscape. The severely fissured appearance of much of Arnhem Land and the Kimberleys testifies to the powerful eroding effect these rivers have wielded over a long period of time. Water, assisted by wind, has sculptured this landscape to produce fantastically weathered features, including mesas, caves, overhangs, cliffs, gorges, waterfalls and stoney outcrops. Spectacular examples of river erosion include the Katherine Gorge and the Jim Jim Falls in Arnhem Land and the Kimberley region's Geikie Gorge, a channel cut by the Fitzroy River through a gigantic limestone coral reef, formed 360 million years ago and since marooned 200 kilometres inland. River erosion is most violent during the wet season.

Both plateau areas are fairly inaccessible and some parts remain unexplored. The Kimberley region is essentially a knot of mountain ranges and plateau areas, interspersed with narrow valleys, and is partly hemmed in by the Fitzroy River to the south west and the Ord River to the east. The average altitude of this landscape is 600 metres, broken by occasional peaks such as Mount Hann (777 metres) and Mount Ord (936 metres). The Napier and King Leopard Ranges run through the western section, roughly parallel to the Fitzroy River, and

The **keelback** or freshwater snake (*Amphiesma mairii*) is found in the vicinity of freshwater streams, lagoons and billabongs from Queensland to Arnhem Land and along the northern coast to the north west corner of Western Australia. Although sometimes encountered in the day it is mostly active at night, especially during the summer. Frogs provide this semi-aquatic snake's chief source of food, but it sometimes eats small fish. Most keelbacks average 75 cm in length, although some have been recorded as reaching 1 m. The keelback is not poisonous and its only defence, other than flight, is to emit a foul-smelling liquid.

the Durack Range sweeps up from the south along the eastern side. Precipitous cliffs up to 250 metres high mark the plateau escarpment along parts of the seaward side. Elsewhere the Kimberleys reach the sea as a series of drowned river valleys. Arnhem Land has a similar profile, being extremely rugged, but is smaller in area and generally no more than 450 metres above sea level. Broken plateau country may also be found in the centre of the region west of the Victoria River, although this terrain is generally lower in altitude than the Kimberleys or Arnhem Land. Closer to the coast and along the major river basins the topography is much lower.

The region's climate is tropical and divided into two seasons, the wet and the dry. Rain is heaviest in the wet, which lasts from November to March with most of the precipitation occurring between December and February, when the north west monsoon crosses the coast. A rainfall of 1500 mm in Arnhem Land and slightly less elsewhere along the coast is common. Further inland the effects of the monsoon are less evident and steadily diminish with distance from the coast. During this season the coast is periodically battered by cyclones which can cause extensive damage, especially if accompanied by the unusually high tides that occur in the region. By April the wet has petered out and during the following dry season, which lasts until October, very little rain falls causing the landscape to become increasingly parched and dusty. Temperatures remain high throughout the year, averaging 32 degrees Celsius in January and 19 degrees Celsius in July, with extremes even greater inland.

Vegetation is tropical and, as can be expected, most luxuriant near the coast in the high rainfall areas. Dense mangrove forests crowd many of the saline estuaries, inlets and creeks that border the Arafura Sea and also to a lesser extent the estuaries of the rivers draining to the Timor Sea from the Kimberleys.

Native grasses, sedges and herbaceous plants dominate the flora found on the coastal plains north of Arnhem Land. This flat landscape is partly covered with swampland, often supporting large stands of tea trees or paperbarks, and dotted with pockets of monsoonal rain forest. Further west towards the Kimberleys the lowlands are generally dominated by the ubiquitous eucalyptus.

This type of vegetation almost entirely covers the lower river valleys and more accessible terrain of the plateau areas. Eucalyptus remains the main species of tree, although the bottle-shaped baobab tree, instantly recognisable due to its swollen trunk which grows up to two and a half metres in diameter, is fairly common. Various grasses and types of scrub, such as acacias, cover the forest floor. As the climate becomes more arid further inland the forest gradually thins out, steadily being replaced by savannah. Trees become lower in stature and the perennial Mitchell and blue grasses become intermingled with spinifex. Eventually, on the southern fringes of the region a desert vegetation, composed of spinifex and low scrub, takes over. Higher up in the plateau regions

The **olive python** (*Bothrochilus olivaceus*) is distributed right across the coast and hinterland of northern Australia, from Cape York to the Kimberleys, and is also found in the rocky gorges of the Pilbara. It is particularly fond of rocky hills and ranges where it is most likely to be seen near water. Most olive pythons average 2.5 m in length but this specimen has reached 4 m, regarded as the maximum length. They are nocturnal hunters and feed mostly on small mammals, such as rats or even small wallabies, which they asphyxiate with their coils and then swallow whole.

Despite this gecko's chocolate coloured body with its distinctive yellow dots and stripes, the **spotted gecko** (*Oedura gemmata*) was long confused with a similar species and has only recently been described in its own right. It inhabits the rocky escarpments of Arnhem Land and lives in crevices. A nocturnal lizard, the spotted gecko hunts insects and is preyed upon by a wide range of enemies. Other than flight, the gecko's only means of defence is its tail which can be jettisoned to distract enemies. This species grows to 15 cm long.

themselves the thinner soils, less reliable rainfall and areas of bare rock create a less hospitable plant environment and the scrublands give way to spinifex, tussock grass and scrub. Within the most rugged plateau environments lush vegetation is usually confined to narrow river valleys and gorges where there is a better chance of finding deeper, more fertile soil and permanent water. Plant cover in these habitats includes wild fig, pandanus, palms, woolly-butt and stringybark.

Because of the region's exceptionally rugged terrain many areas are virtually inaccessible, making any accurate and comprehensive survey of the indigenous wildlife very difficult. Many species have only been discovered fairly recently. These include the splendid tree frog (Litoria splendida), the carpenter frog (Megistolotis lignarius), the cave gecko (Pseudothecadactylus lindneri) and two pythons, Morelia carinata and M. oenpelliensis. The latter snake is thought to be the species that features in many Aboriginal myths. At present the approximate number of cold-blooded species known in this region includes 120 lizards, 60 fish, 44 snakes, 38 frogs, 4 tortoises and both Australian crocodiles.

Of the two crocodile species found in Australia the freshwater or **Johnston's crocodile** (*Crocodylus johnstoni*) is the smaller, averaging 1.2 m in length. Large male specimens sometimes reach 3 m but this is uncommon. This species is fairly widely distributed over northern Australia, from Cape York to the Kimberleys. Its status as a protected animal has made it an increasingly familiar sight in freshwater lagoons, rivers and billabongs. Although sometimes seen during the day, these creatures are mainly active at night when they forage for fish, frogs, birds, crustaceans, small mammals and reptiles. During the day they remain dormant, lying under foliage or just floating on the water. They have few enemies except in their juvenile and egg stages, with feral pigs in particular being great destroyers of crocodile nests. Females lay about twenty eggs just prior to the wet season, depositing them in sandbanks.

Generally around 60 cm long, the **northern blue-tongue lizard** (*Tiliqua scincoides intermedia*) is the largest member of this family found in Australia. It frequents open woodland and rocky areas in Arnhem Land and the Kimberleys and is diurnal, although it is occasionally encountered on warm nights. Usually, however, it spends the night sheltering in hollow logs or beneath debris. Insects, snails, worms and plant matter, such as berries and flowers, constitute the lizard's diet. Like other blue-tongues, when threatened, this lizard puffs up its body and assumes a U-shape, tucking its legs beneath its body, flicking its tongue and hissing. Blue-tongues are live-bearers and females produce litters of up to twenty-five offspring.

Merten's water monitor (*Varanus mertensi*) may be found in coastal and inland water areas right across the northern part of the continent, from western Cape York to the Kimberleys. It is a fairly common aquatic lizard and may be spied basking in the sun — often lodged in branches above the water — near creeks, billabongs and swamps, especially during the early morning or late afternoon. Although it is fairly bold, and when encountered foraging in picnic areas will stand its ground, this creature usually takes to water quickly if disturbed. It is able to stay submerged, roaming around the bottom in search of fish, frogs or freshwater shrimps and prawns, for several minutes at a time. This close-up shows the position of the lizard's nostrils. Their location, high on the snout, enables the lizard to breathe more easily whilst submerged. When the lizard requires air it merely raises the top of its head to the surface.

The **giant cave gecko** (*Pseudothecadactylus lindneri*) inhabits the more rugged and dissected terrain of Arnhem Land and is most likely to be seen in caves and crevices. A subspecies is found in the Kimberleys. Nocturnal in habit, it spends much of its waking hours foraging for insects, such as the cockroach gripped in this specimen's mouth. Occasionally during the wet season, it ventures outside its usual cave environment in search of food. The adhesive pads or lamellae on its feet (also on the tail) are ideal for running around rock surfaces.

This landscape, located in the western part of **Arnhem Land** towards the Daly River basin, is dominated by scattered dry sclerophyll tree cover and perennial grasses; which have been reduced to virtual stubble by the prolonged dry weather. The threatening storm clouds indicate that the onset of the wet season is imminent and after rain falls this landscape quickly rejuvenates. The best known creature that lives here is the frilled-neck lizard.

This **frilled-neck lizard** (*Chlamydosaurus kingi*) is signalling its alarm by spreading its huge frill in an elaborate display of bluff. The frill normally lies in folds around the neck and shoulders and is erected only when the lizard opens its mouth widely. This strategy is usually very effective in warding off danger but if it proves to be insufficient the lizard menaces the enemy by lashing its tail. If the confrontation develops into an actual fight the lizard can inflict painful bites with its large canine teeth. This particular lizard species is probably one of Australia's most famous reptiles and it can be found throughout much of northern Australia, from southern Queensland to the Kimberleys, via Cape York and Arnhem Land. Frilled-neck lizards normally live in open sclerophyll forest country, where they spend much of their time in trees hunting insects. They are usually active during the day and are particularly fast-moving creatures, when running on their hind legs. Most specimens average 70 cm in length, although they sometimes grow to 90 cm. Males are usually more brightly coloured than females and also have heavier built jaws.

The **two-lined dragon** (*Diporiphora bilineata*) is most often seen in fairly open country in the Kimberleys and Arnhem Land and can easily be observed during daylight hours sitting in an elevated position. This particular dragon is clinging to the summit of a termite mound. Rarely more than 25 cm long, most of the length being tail, this dragon hunts insects and is heavily preyed upon by birds and other predators. Faced by an enemy the dragon's only defensive behaviour is to open its mouth and hiss.

Although many species of Australian dtellas — of which the best known is probably the native house gecko (*Gehyra australis*) — have been described in recent years it seems that many others still await discovery. One of these 'unknowns' is the specimen pictured here, which was located under loose bark on a fallen tree 50 km south of Darwin. Nothing is known yet about its biology.

Australia's largest reptile is the **salt water crocodile** (*Crocodylus porosus*) which averages 2–3 m in length, although specimens over 6 m have been spotted, especially now that the species is protected. They inhabit the coastal regions of northern and north eastern Australia and, despite their name, live in freshwater environments as well as estuaries and salt water swamps. These crocodiles mostly hunt at night and may be seen during the day, basking in the sun on riverbanks, especially during the winter months. They are very aggressive hunters, feeding on fish, waterfowl and animals, such as kangaroos and wallabies, which they grab at waterholes. On occasion they become maneaters. After making a kill they stash their prey under water, lodged in snags or riverbanks, and only return to feed when the carcase has begun to decompose. Because of their size, prowess and protected status they have few, if any, enemies once they reach adulthood. Juveniles are more vulnerable and may be taken by feral pigs, goannas and birds of prey. Females lay batches of up to sixty eggs in a carefully constructed nest. They guard the nest and assist the young when hatching occurs.

Luxuriant tangles of **pandanus**, opposite, or screw palms commonly crowd watercourses throughout Australia's tropical north. These plants are growing beside the Katherine River in Arnhem Land. The pandanus has long sword-like leaves, with spikes at regular intervals along the edges, and a huge brush-like root system which, when intertwined with those of other plants, offers protection for many animals, from rainbow fish to freshwater crocodiles.

Often mistaken for a snake, **Burton's legless lizard** (*Lialis burtonis*) belongs to a lizard family that is endemic to Australia and New Guinea. It is the most commonly encountered representative of this family and is found throughout most of continental Australia in all forms of ground or low-lying vegetative habitats. Although mostly active at night it is also seen fairly often during the late afternoon or early morning. Coloration and pattern vary widely in different habitats; the lizard shown here is typical of those found near Darwin. Other specimens range in colour from brown through black to red; striped lizards of this species are also known to exist. Growing up to 70 cm in length, this lizard hunts skinks and geckos. Although it is able to drop its tail if grabbed there, the legless lizard's best form of defence is its effective camouflage.

Australia's most widely distributed venomous snake is the **king brown snake** (*Pseudechis australis*), also called the mulga snake. It is found in a wide number of habitats, from desert to open forest, and is especially common in semi-arid territory. King browns are most active at night in this region, although they may sometimes be seen in the early morning or late afternoon. Fully grown specimens sometimes reach lengths of 2.5 m and because of their size and toxicity have few enemies, apart from feral pigs. They feed mainly on small mammals and are also known to eat other snakes, including venomous species. Coloration varies with habitat, ranging from reddish to olive-brown to a copper colour. This specimen is typical of the king browns found near Darwin. Along with the death adder, taipan, tiger snake and common brown snake, this species is one of the most venomous creatures found in Australia.

Although long known to exist in New Guinea, the pig-nosed or **pitted-shelled turtle** (*Carettochelys insculpta*), below, was first discovered in Australia in 1969 and has since been spotted in the Daly, Alligator and Victoria Rivers in the Northern Territory. Its nests with eggs have only been discovered recently in Australia and unlike tortoise eggs, which are elongated, they are spherical in shape. The late discovery of this species in Australia is something of a mystery as ancient cave paintings indicate that Aborigines have long known of its existence. This fairly heavy-bodied turtle lives in lagoons, rivers and estuaries and eats small fish, snails, pandanus fruits and water plants. Unlike the other species of Australian tortoises, which retract their heads by twisting their necks sideways, this creature is able to pull its head straight back underneath its shell when threatened, like most other turtles and tortoises elsewhere. The pig-nosed turtle grows to 70 cm in length.

Named after the high black dorsal fin found on males, the black mast or **strawman fish** (*Quirichtys straminous*), opposite, belongs to the same family as the hardyhead species. It is found in northern rivers with hard alkaline water, such as the Daly and Victoria Rivers, and may be seen swimming amongst pandanus and exposed tree roots, searching for algae and similar microscopic organisms. Because of its small size — about 7.5 cm from head to tail — it is hunted by most larger fish.

Yellow Waters Lagoon is fairly typical of the large lagoons that dot the plains crossed by the rivers draining Arnhem Land. The water is partly covered by a variety of aquatic plants and grasses and is populated by stands of paperbarks or tea trees, some of which may be seen on the horizon. Pandanus palms grow in the foreground. These lagoons are at their most extensive in the wet season when large areas of lowland are inundated by rivers which flood and coalesce. Abundant fish life makes these lagoons a favourite haunt of salt water crocodiles which sometimes travel overland to the lagoons from river estuaries.

The **saratoga** (*Scleropages jardini*), like its cousin from Queensland's Fitzroy River system, the spotted barramundi, is a large fish, growing up to 90 cm long. It is randomly distributed throughout freshwater systems across northern Australia, from the Jardine River on the western tip of Cape York to the Adelaide River in the Northern Territory. Oddly, it only appears in some of the waterways in this region and the reason for its erratic distribution is unknown. Commonly encountered amongst lily leaves and stalks in quiet backwaters and lagoons, saratoga are essentially mid-water feeders but will take anything they can swallow. They are extremely territorial fish and will defend their particular hunting grounds against intruders from the same species. This aggressive behaviour is less obvious during the breeding season, when males look for mates. Saratoga are mouth-breeders and lay a limited number of large eggs which are kept in the mother's mouth. This protective behaviour persists through the juvenile stage although young fish do occasionally leave their mothers to feed. Eventually, when they have reached 4 cm or thereabouts in length, the juveniles start to fend for themselves. Growth is rapid in the first year and saratoga reach adulthood at four or five years of age.

Scotch Creek is part of the Mary River system just east of Darwin off the Stuart Highway. The lower reaches of this river flow across flat swampy plains and the waterway is hedged in on either bank by dense tropical vegetation, including paperbarks, freshwater mangroves, pandanus and a large variety of aquatic plants. Similar habitats may be found along many other rivers in this region. The environment is teeming with wildlife, including salt water crocodiles, goannas, green and brown tree snakes, water pythons, carpet snakes, whip snakes and numerous fish, such as silver barramundi, blue-eyes, glass perch, catfish, saratoga, archer fish and many other types.

The **archer fish** (*Toxotes chatareus*) is one of the more common, and most unusual, fish found throughout the river systems of northern Australia. It favours a variety of habitats, from the brackish water of mangrove swamps and estuaries to freshwater lagoons and billabongs and the upper reaches of streams. Generally active during the day, the archer fish is very inquisitive and will swim along the surface to observe objects above the water that attract its attention. Fully grown specimens average 30 cm in length and juveniles sometimes grow up to 15 cm in their first year if food supplies are plentiful. They are normally a silvery-yellow colour with dark blotches on the back and upper part of the body and a dorsal fin set well back on the spine. Archer fish are most famous for their unique method of catching food — literally knocking insects from overhanging branches with a thin jet of water. They are able to propel jets of water from their mouths by compressing their gills. This causes a thin stream of water to issue from a nozzle, formed by a groove in the fish's palate, in the centre of the closed mouth. The resulting jet of water can be fired at targets up to 1.5 m away — a feat which has earnt this species its alternative name of rifle fish. This fish has successfully downed a cockroach. Besides catching prey in this spectacular manner, archer fish also eat smaller fish which they take in a more conventional fashion beneath the surface.

Although the **black catfish** (*Neosilurus ater*) is widely distributed throughout northern Australia, because of confusion with similar species of catfish it is difficult to determine the range of this species with great accuracy. However, it is known that they are commonly found in a number of freshwater habitats, from lagoons and billabongs to flowing streams. Mostly active at night, but sometimes seen in the early morning or late afternoon, black catfish feed chiefly on worms and small crustaceans. Their coloration is normally very dark and black spots may be seen on one or both sides of the fish. Adults grow to about 60 cm long.

The existence of **Rendahl's catfish** (*Copidoglanis rendahli*) has been recognised in New Guinea for some time but this species was only discovered in Australia recently. It is found in river systems between Cape York and the Northern Territory and possibly in the Kimberleys. Small creeks, lagoons and billabongs are its most likely habitats. After spending most of the daytime lurking amidst aquatic plants and rocks this catfish emerges at night to eat worms and crustaceans. Up to 20 cm long, its enemies include water birds and larger fish.

The **red-tailed rainbow fish** (*Melanotaenia splendida australis*), left, is widely distributed throughout much of north western Australia. This particular form, with its bright red tail, is found in the upper reaches of the South Alligator River in Arnhem Land and grows up to 12 cm long. It prefers soft clear acidic water in habitats distinguished by dense aquatic vegetation including copious leaf litter on the bottom of the stream.

This particular colour form of the **banded rainbow fish** (*Melanotaenia trifasciata*) is found inhabiting clear water streams, such as the Giddy River in Arnhem Land. Other forms with different coloration are found elsewhere. The banded rainbow is a very active species and mostly feeds on small crustaceans and insect larvae. An even more colourful form of this species is found in the Goyder River, also in Arnhem Land. The Goyder River form has bright red fins and red stripes on top of a blue body. Both forms reach 12 cm in length.

The **dainty blue eye fish** (*Pseudomugil tennelus*) was first discovered by an American museum expedition in 1948 but the species remained fairly mysterious until it was described scientifically in 1964. Live specimens were not acquired until recently and this is the first photograph ever taken of a male. The dainty blue eye lives amongst dense aquatic plant growth in lagoons in the northern part of the Northern Territory.

Usually seen alongside streams and swamps or in open forests and rocky habitats, the **northern tree frog** (*Litoria wotjulumensis*) is distributed throughout Arnhem Land and the Kimberleys. Up to 7 cm long, this ground-dwelling tree frog is nocturnal and hunts insects.

The **red-eyed tree frog** (*Litoria rothi*) is usually found near the coast or close to large river systems in the northern part of Australia, from Cape York, through the Gulf Country and Arnhem Land, to the Kimberleys. Typical habitats include lagoons and creeks. It is nocturnal and hunts all kinds of arthropods. Coloration varies slightly with habitat, but all species feature bright yellow markings on the upper thighs that act as warning colours whenever the frog jumps. This amphibian has a very distinctive trilling call and grows to about 7 cm in length.

Cyclorana australis is closely related to the northern burrowing frog, *Cyclorana novahollandiae*, and is found from north west Queensland throughout most of the Northern Territory to north west Western Australia. Like its cousin this species spends most of its time hidden underground and is usually only seen in the wet season. In order to survive its subterranean existence this frog stores water like other members of the genus. Its call is a loud and penetrating honk honk, which carries over a considerable distance. Due to the transitory nature of this frog's breeding grounds — they only exist after rain — the development of larvae into tadpoles and frogs is rapid and largely unknown.

Usually glimpsed in hills, ranges and rocky outcrops where it seeks sanctuary in crevices and caves and beneath large boulders, the **rock goanna** (*Varanus glebopalma*) is found throughout the Kimberleys, Arnhem Land and the western section of the Gulf Country. Although quite common, this goanna is exceptionally shy and is rarely seen. It is most likely to be encountered at twilight when it hunts all kinds of skinks, birds and small mammals. This particular specimen was found sleeping on a rock ledge inside a cave and is displaying typical goanna defensive behaviour, erecting the front of its body and puffing out its throat. However, these tactics are only employed when the goanna is cornered, as this species is able to scurry away at a very fast rate. Because of its speed it has few real enemies apart from some snakes. The goanna's coloration varies with different habitats and this species grows up to 90 cm long, most of which is tail.

This small and rocky valley is typical of many similar habitats found along the escarpments that mark the east and western boundaries of the Kimberley plateau. During the wet season these valleys are filled with swift streams that pour out from the plateau interior, en route to broad river valleys such as the one in the distance. For the rest of the year these places are dry and, compared to the lush habitats found closer to the coast, relatively barren. Drought resistant plants, such as the acacias, eucalypts and spinifex grasses seen here, testify to the area's intermittent drainage and seasonal rainfall. Local animals include spiny-tailed and velvet geckos, carpenter frogs, rock goannas and splendid tree frogs which live in the many caves that honeycomb this terrain.

Although the **brown tree snake** (*Boiga irregularis*) puts on a very aggressive display once it is aroused it usually tries to evade its enemies. This specimen has coiled itself up into a series of S-shaped loops, preparatory to striking, which it does accurately and repeatedly. It is one of Australia's few rear-fanged snakes, with its fangs located at the back of the jaw, and although venomous is not considered dangerous to humans. Various colour forms of the brown tree snake inhabit the coastal regions of eastern and northern Australia, from mid New South Wales to the Kimberleys. This red and white banded form is sometimes called the night tiger and is found in Arnhem Land and the Kimberleys. Normally about 1.5 m long and fond of rocky outcrops near water, this snake hunts frogs and mammals at night.

These **giant baobab** trees are a fairly common sight on the flat savannah plains of the Kimberley country. They usually vary considerably in shape and, along with eucalypts and acacias, provide the dominant tree cover in this type of terrain. Perennial Mitchell and spinifex grasses occupy the lower plant storey. The indigenous fauna includes black-headed pythons and knob-tailed geckos.

This **black-headed python** (*Aspidites melanocephalus*) includes venomous snakes among its victims. Small mammals, ground birds and other reptiles are also squeezed to death. Although sometimes spotted during the early morning or late afternoon this constrictor is most likely to be encountered at night. It is fairly adaptable and can be found in coastal forests and more inland habitats located throughout northern Australia, from Queensland to the Kimberleys. Because of its size — fully grown adults reach at least 3 m in length — this python has few enemies. Shorter specimens are sometimes confused with the poisonous tiger snake, although that snake is only found in the south of Australia.

The rather grotesque-looking **rough knob-tailed gecko** (*Nephrurus asper*) stalks intended victims with cat-like stealth, waving its tiny tail from left to right whilst fixedly staring at the prey in an apparent bid at mesmerism. If alarmed the gecko raises its body on all four legs, lifts its tail and then advances on the baffled enemy while making barking sounds. This bluffing strategy is given some extra credence by the gecko's length, at 20 cm long it is one of the largest members of the gecko family in Australia. It normally lives in savannah country or rocky habitats throughout northern Australia and eats insects and spiders.

Bushy pandanus plants and various tropical grasses crowd this slow-moving stream in the Kimberley Region. Similar habitats exist wherever there is permanent water.

The spiny-tailed gecko (*Diplodactylus ciliaris*) is found throughout much of northern Australia, from the Kimberleys to Arnhem Land. It is often seen crossing roads at night and is less frequently observed crouching motionless on the bark of trees, where its camouflage is most effective. It is named after the soft spines on its tail which can eject a sticky, irritating substance, suitable for spraying in the eyes or mouths of predators, such as snakes, carnivores and birds. Like all geckos, the spiny-tailed gecko is nocturnal and eats small insects.

This distinctively patterned **velvet gecko** (*Oedura marmorata*) is from Manning Creek in the Kimberleys. The species is quite common throughout much of northern, central and western Australia and is generally found in rocky habitats, such as crevices and gorges, often spending the day inside caves and only emerging at night to hunt. Averaging 15 cm in length when fully grown, the velvet gecko eats insects and spiders and is preyed upon by birds and snakes. It is an extremely agile lizard, able to jump considerable distances, and can run very fast.

Juvenile **red-faced tortoises** clearly display the red strip that gives this species its name. During breeding times female tortoises dig nests high above the water level, burying ten to twenty eggs that usually take three months to hatch. The juveniles are about 2.5 m long at birth and are preyed upon by birds and crocodiles. Adults have few enemies and will bite if molested. This specimen is about one year old and measures approximately 10 cm in length.

Red-faced or **boof-headed tortoises** (*Emydura australis*), left, reach a shell length of 30 cm and are found fairly readily in watercourses throughout the Kimberley Region, where they can be observed swimming about in search of mussels, shrimps and other food. They are usually a blackish, dirty brown colour and their oval-shaped shells are often given a greenish tinge by algae growth.

Notoscincus ornatus wotjulum is a small skink about 10 cm long that generally lives in leaf litter beneath trees, such as the baobab, in the west Kimberley region. Very little is known about these lizards. They feed on small invertebrates, provide food for birds, larger lizards and snakes and have no common name.

One of Australia's most spectacular dragons, **Gilbert's water dragon** (*Lophognathus gilberti*) grows up to 45 cm long. It is found between Queensland's Gulf Country and Western Australia and is usually seen in trees, although specimens have been spotted on the beaches of North West Cape. This diurnal lizard mainly eats insects and is eaten in turn by birds of prey. Females are usually a grey-brown colour and males develop darker markings in the breeding season between September and December. These lizards are fairly territorial and usually live in colonies composed of several males with their families.

Mitchell's water goanna (*Varanus mitchelli*) is one of the three Australian goannas specially adapted to aquatic life. The others are the mangrove monitor (*V. indicus*) and Merten's water monitor (*V. mertensi*). All species have a flattened tail which acts as a rudder when the goanna is swimming. This species is found beside watercourses in Arnhem Land and the Kimberleys, particularly around the Ord River Irrigation Project where it may be spotted sitting on dams or bridges. However, if approached it will quickly dart towards the water and disappear. Frogs form a major part of this diurnal lizard's diet, a food that causes them to be infested with parasites. Most specimens grow to 70 cm long.

The slender pincer-like jaws of the **long tom** (*Strongylura kreffti*) are armed with rows of needle-like teeth which make this fish a formidable hunter. The long tom preys chiefly on other fish which it grabs with a swift sideways swipe, turning its prey around to be eaten head-first. Juvenile long toms usually hunt in schools, becoming solitary predators once they have grown to their full adult length of 75 cm. Although this species is widely distributed throughout northern Australia and easily observed swimming near the surface, surprisingly little is known about its biology.

Almost all the streams in the Kimberley Region are intermittent. During winter they dry up, leaving small isolated pools along the river beds to support plant and animal life. Sometimes, if the winter is particularly dry and the rains start late, these waterholes evaporate altogether. Seven species of fish were found in this one, including rainbow fish, catfish and grunter. Gilbert's water dragons inhabit the paperbarks and pandanus that grow along the river bank.

Surprisingly, considering its status as one of Australia's largest frogs, growing to about 15 cm long, *Litoria splendida* was only discovered in the mid 1970s. So far it is only known to exist in the Kimberley Region where it inhabits trees, rocky areas and moist places in and around houses. During the dry season this frog preserves moisture by clustering in caves with others of the same species. Adults have no known enemies and, as they are closely related to the common tree frog, *Litoria caerulea*, it is assumed that they also eject a poisonous fluid from glands on the head when menaced. They feed chiefly on insects, such as cockroaches and beetles, and are usually bright green with yellow spots and pinkish thighs.

The **western sooty grunter** (*Hephaestus jenkinsi*) is found in streams and clear pools in rocky habitats throughout the Kimberleys. It is often seen swimming amongst submerged tree roots. An attractively coloured species, it is able to change its appearance to suit different environments and feeds on smaller fish, molluscs and crustaceans. This species grows to about 50 cm long.

The **Kimberley sharp-nosed grunter** (*Syncomistes kimberleyensis*), opposite, is similar in appearance and habits to the western sooty grunter but even less is known about this species. It occupies rivers throughout the Kimberley Region and is active during the day when it eats plant matter, algae, insect larvae and small crustaceans. So far only juveniles have been caught and the length of the adult fish is unknown.

The **slender rainbow fish** (*Melanotaenia gracilis*), above, found in habitats such as the Drysdale River in the Kimberleys, is one of the prettiest and most delicately coloured members of the rainbow family. It favours clear water where there is a sandy bottom and can be seen foraging for food during the daytime. Besides being eaten by larger fish it is also part of the freshwater crocodile's diet. This species grows to 7.5 cm in length.

The **long-snouted water dragon** (*Lophognathus longirostris*) is a small arboreal lizard most often seen nestling in tree branches or foraging amongst ground debris beside watercourses in arid regions from the centre of the continent to the Western Australian coast. If approached by a potential enemy this dragon usually tries to escape, erecting itself on its hind legs, lifting its tail in an arc and dashing off at high speed. It grows to about 35 cm in length, most of which is tail, and chiefly feeds on insects.

For much of its length and for most of the year Western Australia's **Fortescue River** is subterranean. However, permanent surface water is found in large ponds, such as this one located near Millstream Station. Despite the high rate of evaporation in the summer the water level is replenished from the underground reservoir and remains fairly constant. River red gums and paperbarks grow along the bank and reeds and lilies populate the shallows. Numerous aquatic plants support a thriving food chain and fish species in this habitat include fresh water herrings, several types of grunter, catfish and gudgeons. Among the reptiles found in the vicinity are Gould's goanas, black-headed monitors and the long-necked tortoise.

4
THE PILBARA

LOCATED IN THE NORTH WESTerly corner of Western Australia and flanked either by the Indian Ocean or by desolate deserts and salt lakes, this drainage region incorporates those rivers which fan out towards the sea from the Pilbara, Hamersley and Yilgarn Plateaux. Ranging from north to south the major rivers and their tributaries are the De Grey, Oakover, Shaw, Coongan, Fortescue, Ashburton, Lyons, Gascoyne, Wooramel and Murchison. Due to seasonal and generally unreliable rainfall, especially in the tropical north where precipitation is largely caused by erratic cyclones, all are intermittent in flow. Most flow for only a few months each year — those situated in the north running in summer and those in the south flowing during winter — and their annual discharge is variable. For much of the year these rivers hardly merit their name as they constitute little more than a trickle of water. Often they dry up entirely along large sections of their course, leaving what permanent water remains stranded in isolated pools. Sometimes, as with the Fortescue River, they even disappear completely, retiring beneath the ground to form subterranean rivers.

With the advent of the wet season which occurs in either summer or winter depending on area, many of the rivers briefly transform themselves into raging torrents, especially where they roar out of the plateau country. Some flood, although the results are nowhere near as extensive as they are in the northern monsoon regions. Although many rivers rise at the edge of the plateaux areas and run directly towards the sea others are hundreds of kilometres long and travel towards the coast, collecting tributaries as they go, by fairly circuitous routes. Besides the Gascoyne River, which at 820 kilometres from start to finish is the state's longest waterway, other lengthy rivers include the Fortescue (600 km), the Ashburton (650 km), and the Murchison (700 km).

Western Australia's Wittenoom Range, opposite, features the mesas or flat-topped mountains and severely eroded escarpments typical of much of the Pilbara region. This environment receives irregular seasonal rainfall and for most of the year is fairly arid, becoming incredibly hot during the summer. The vegetation reflects this harsh climate and the plants shown here — spinifex grasses, low acacias and eucalypts — are all able to cope with drought conditions. The most lush vegetation is found surrounding permanent sources of water usually contained in isolated pools. These pools feature a good deal of plant and microscopic aquatic life which provides food for a thriving fish population, including species such as catfish, striped grunter and spangled gudgeons. Reptiles include Gould's goanna, the spiny-tailed goanna, the diner-plate tortoise and the striped gecko, which is particularly suited to life in spinifex clumps.

The region's basic relief is primarily composed of lowland, occupying the coastal plains or the major river valleys, and plateaux, occasionally broken by peaks like Mount Bruce (1235 metres) or mountain ranges such as the Hamersleys. The plateau regions sprawl across much of the land area and serve as buffers between the arid interior, populated by the Great Sandy and Gibson Deserts, and the coastal lowlands. Generally no more than 300 metres above sea level, topographically the Pilbara, Hamersley and Yilgarn plateaux have much in common. Predominantly flat and barren or else gently rolling for much of their extent, the plateau surfaces have been torn apart at intervals by folding and faulting of the earth's crust and subsequently much shaped by erosion. Rain and wind-blown sand have carved and buffeted this surface with spectacular effect, creating magnificent gorges such as the Yampire, Hamersley and Wittenoom. Many of these chasms have sheer walls faced with horizontal bands of red, brown, blue or purple rock. Except in the east, where the terrain becomes steadily engulfed by sand as it merges with the deserts of the Great Western Plateau, this high country is normally demarcated by precipitous escarpments that rise from the surrounding plains. These ramparts are sometimes indented by rugged, steep-sided gullies divided from one another by long spurs covered with scree or rock debris. In some places, noticeably in the south, hills rising above the general plateaux surface have had their walls severely eroded so that the hilltops overhang the sides, forming 'breakaways'.

The coastal plains are generally fairly narrow. In the north they are fringed with swamplands but the plains are normally fairly dry. Between Exmouth Gulf and the Wooramel River they are widely covered with sand hills and dunes. Lake McLeod, a huge salt lake of the type normally encountered in the interior, lies on this plain between Exmouth Gulf and Shark Bay, the only significant indentations on an otherwise regular coastline.

Climatically the region is variously affected by both tropical and temperate weather systems. The north lies within the zone of tropical influence and receives summer rainfall between December and May. Total average 200 mm, although heavier falls up to 350 mm are recorded in the Pilbara region. Rainfall tends to decrease inland and further south and is extremely unreliable, both in terms of its frequency and its quantity. This is because precipitation is largely associated with cyclones which, accompanied by high winds and torrential rainfall, sweep across the coast, sometimes with devastating effect, throughout the summer. Some seasons receive more cyclones than others, hence the discrepancy in annual rainfall totals. This reliance on cyclonic rainfall has produced a weather pattern that tends to alternate between flood and drought, a cycle that strongly determines plant cover. The southern zone is transitional in climate, with tropical influences being experienced in the north and temperate ones in the south. Most rain falls between March and August and averages 200 mm per year. Temperatures are high throughout both climate zones and the region has a January mean of 30 degrees Celsius and a July mean of 19 degrees Celsius. However, summer temperatures frequently produce prolonged heat waves and Marble Bar, in the Pilbara, has the Australian record for the most consecutive days over 37.8 degrees Celsius.

Unreliable seasonal rainfall combined with rugged terrain has produced a vegetation cover that is strongly drought resistant.

The long-necked or **dinner plate tortoise** (*Chelodina steindachneri*), opposite, is endemic to the north western corner of Australia where is can be found in desert-type intermittent rivers from the De Grey south to the Irwin River. It is usually seen in very isolated pools and has been known to survive in mud for a long time, perhaps several years. Otherwise very little is known about this reptile which grows up to 20 cm in length.

The **striped grunter** (*Amniataba percoides*) is one of the most widely distributed and common species of this family found in Australia. It thrives in virtually all freshwater environments and is commonly found throughout this region. The striped grunter is active during daylight and will eat anything it can overpower while in turn it provides food for larger fish and water birds. Usually between 10–15 cm long when fully grown, this species possesses a consistent coloration despite its wide distribution.

Although mangroves line the estuaries and bays along part of the coast as far west as the Ashburton River mouth, the flora in this region is not predominantly tropical in appearance. The low-lying areas to the north are generally covered with an assortment of perennial and ephemeral grasses, heath and scrub. Further south the sand-hill country is host to a variety of hummock grasses and low scrub or woodland, mostly composed of acacias, banksias, grevilleas and hakeas. After rain this generally rather drab and uninteresting landscape is transformed into vivid carpets of colour as numerous wild flowers produce their blossoms. Heavier timber, notably eucalypts such as ghost or river red gums, bloodwoods or coolibahs, grows along watercourses and river valleys throughout the region wherever there is a guaranteed supply of water. Ferns are sometimes found growing in deep shady gorges where water seeps from rocks. Where water is in the habit of regularly drying up the prevailing plant cover consists of hummock grasses and scattered scrub.

This vegetation extends into the plateau country except where scree or bare rock prohibit growth. The most common hummock grasses found growing throughout the region are the various species of Triodia, popularly known as spinifex. These plants are ideally suited to arid terrain, such as the sand-hill country to the south or the plateau surfaces. They are perennial summer growers composed of dense hummocks of yellowy-white prickles, topped by stalks up to two metres high. Spinifex grasses are often found dwelling as an understorey beneath mulga scrub, one of the chief species of acacia encountered in this region. The drought resistant mulga grows up to ten metres tall and features a spindly trunk and branches that support a sparse crown of greyish-coloured leaves.

The predominantly arid and desert type terrain associated with this region is too harsh for many cold-blooded species. The region's twelve fish species are forced to spend much of their existence surviving long dry spells in permanent water holes. This condition limits the distribution of many fish species and two — the blind cave fish — are found only in isolated sinkholes on the North West Cape. Similarly the climatic conditions have restricted the local tortoise population to a single species, the long-neck or dinner plate tortoise (Chelodina steindachneri), while most of the fourteen species of frog found here are burrowing creatures. Snakes and lizards appear in far greater numbers, ranging from thirty to 144 species respectively.

Australia's largest lizard is the **perentie** (*Varanus giganteus*) and specimens over 2 m in length have been reported. However, most perenties are far smaller, averaging 1.6 m. Distributed throughout Australia's arid interior, from western Queensland to the Pilbara, this ground-dwelling lizard usually prefers rocky terrain where it lives in crevices. It is not restricted to this habitat though and has been encountered foraging in sand dune country. The perentie hunts during the day and small mammals, such as rabbits, form a major part of its diet. Because of its size the perentie has few enemies. If attacked it defends itself in the same way as other goannas and when necessary can cause considerable damage with a blow from its large and powerful tail. Juveniles, however, try to evade detection and lead secretive lives.

The **striped gecko** (*Diplocactylus taeniatus*) has made the spinifex clump its home. Ideally camouflaged for such a habitat, this lizard spends each day clinging motionless to a grass stem, pointing its head down and keeping the body straight. During the night it forages throughout the spinifex clump and around the surrounding terrain in search of insects and spiders. If disturbed from its daytime repose by a goanna or snake the gecko will suddenly open its mouth wide, disclosing a bright mauve interior. This effect is sometimes dramatic enough to frighten off predators, but if it fails to do so the gecko ejects an unpleasant fluid at its enemy from glands located in the tail. This species is found in the Pilbara and the desert regions to the east and is not particularly common. It grows up to 8 cm in length.

This rugged **gibber plain** is covered with scattered hummocks of spinifex grasses. Wind erosion has removed most of the topsoil, leaving a legacy of small pebbles and stones, and rainfall is infrequent and scanty. Despite the aridity and harshness of this environment, where temperatures can soar during the day and plunge at night, it is populated by numerous species of reptiles. At least thirteen species of gecko, such as the spiny-tailed gecko and the striped gecko, live in the spinifex. They provide food for larger lizards like Gould's goannas, spiny-tailed goannas and Burton's legless lizards and snakes such as the desert death adder.

The striking looking **spiny-tailed gecko** (*Diplodactylus strophurus*) is usually seen on the road at night when it hunts insects and other invertebrates and tries to evade predators. It is less likely to be encountered during the day when it hides in spinifex clumps. Little is known about its defensive behaviour, although it is possible that the unusual scale arrangements on its tail might be designed to scare off enemies. The tail scales are separated by bands of yellow white connective tissue which show up against its predominantly grey-white coloration whenever the tail moves. Adult geckos grow to about 10 cm long.

The **desert death adder** (*Acanthophis pyrrhus*) lives in arid desert conditions throughout continental Australia except Victoria. It is distinguished from the other two species of death adders by its rough scales, vivid coloration and more slender body shape. Extremely venomous it can strike rapidly and grows to about 70 cm in length.

The **hooded scaley-foot** (*Pygopus nigriceps*), opposite is a large legless lizard about 45 cm long. Found in desert habitats throughout Australia it hunts at night, mainly catching invertebrates. When attacked it sheds its tail, hoping to escape in the confusion. Alternatively, it erects the front part of the body and strikes wildly at its enemy, which has difficulty deciding whether it is confronted by a snake or a lizard.

Red sand dune country is the predominant terrain in the southern part of this region and also throughout much of central Australia. Most dunes rise to about 10 m and are held in place by low acacia, hakea, grevillea and casuarina scrub and spinifex grasses. Because of the intense temperatures most of the animal inhabitants, such as the desert death adder, mountain devil, knob-tailed gecko and hooded scaley-foot, are nocturnal. The **red desert dragon** (*Ctenophorus rubens*), right, is found between North West Cape and the Pilbara. It mostly feeds on insects and can sometimes be observed darting around quickly, bobbing its head and moving a front leg in a curious waving motion before resuming its search for food. This dragon is extremely nervous and difficult to approach. It grows to 30 cm long.

At 20 cm in length the **short-tailed goanna** (*Varanus brevicauda*) is the smallest representative of this family found in the old world. It lives in sandy and gravelly spinifex country in north west Australia and the Northern Territory and has recently been found in western Queensland. Although sometimes encountered wandering around on warm nights, this lizard is an active burrower and prefers to spend the nights underground. It emerges during the day to hunt insects and lizards. This pair is fighting over a gecko. The goanna on the left is male and its antagonist is female.

Very few people have ever bred this species in captivity. This pair of short-tailed goannas, opposite, was kept in a glass case with 15 cm of sand in the bottom. After hibernation — simulated by turning off the heating and lighting for a month — they emerged from their burrow and immediately began to mate. The male chased the female for several hours as part of the goannas' mating procedure before attempting to copulate. This pair eventually copulated several times. The female laid eight eggs in three batches over a six-week period. Hatching took from six to twelve weeks.

This juvenile, inset top right, took a whole day to emerge from its egg. After slitting the egg open with its egg tooth — seen on the goanna's snout — the juvenile spent twelve hours resting with only its head breaking through the shell before emerging completely.

This newly-hatched juvenile, inset lower right, is only 7.8 cm long. Within two years it had reached its full adult size. Short-tailed goannas are so named because of their equal tail to head and body ratio. Most other species have tails that are significantly longer than the combined head and body length. Short-tailed goannas are fairly variable in their colouring, ranging from fawn to pinkish or yellowish brown.

Although this valley in the North West Cape region contains water during the erratic summer rainy season, throughout the remainder of the year it is almost completely dry. Any water that might be expected to lie in permanent pools during the dry season seeps away through the porous limestone that comprises the dominant rock type in this range.

Often seen standing in fairly prominent vantage positions beside roads or on logs and rocks, the central **netted dragon** (*Ctenophorus nuchalis*) is found in arid country throughout central and western Australia. This specimen is standing beside a limestone hole ready to disappear if threatened. Whenever this species becomes alarmed or too hot their first instinct is to seek cover. It is active during the day, particularly in the early morning or late afternoon, when it hunts insects. Goannas, foxes and birds of prey are the dragon's main enemy. Colour varies with habitat, ranging from reddish to orange brown, and this specimen is typical of those found in the North West Cape Range.

This sinkhole, connected to several others by water-filled tunnels, is located in the arid coastal plains south-west of Exmouth on North West Cape. The desolate looking landscape is composed of porous limestone and covered by scattered salt bush and spinifex plants. Several blind creatures, including the blind gudgeon (*Milyaringa veritas*) and the blind eel, live in the subterranean tunnels. Whereas the blind eel is very rare — only four specimens have ever been found — the blind gudgeon is comparatively common and can be readily observed in the brackish water of these sinkholes, especially at night with the aid of torches.

The reason why the **blind gudgeon** came to be occupying this particular habitat is uncertain, as 5000 years ago this plain lay beneath the sea and this does not seem long enough for the species to develop from scratch. Instead, it is thought that the blind gudgeon evolved in caves inside the North West Cape Range, subsequently expanding its habitat by moving into the subterranean water courses of the coastal plain. Because very few blind gudgeon have been studied in captivity very little is known about this fish's habits and biology. Like many other blind fish species found elsewhere in the world, blindness does not appear to be a handicap for the blind gudgeon which is well adapted to life in the dark, having many sensory organs which are sensitive to touch around the snout and head. As soon as these organs make contact with food, such as mosquito larvae, the blind gudgeon snaps at its prey with great accuracy. Since other gudgeon species lay clutches of adhesive eggs that are guarded by males it would be interesting to learn how this particular species goes about breeding. Unfortunately nothing is known about this aspect of the blind gudgeon's existence.

Due to low rainfall and poor soils much of this region is covered by stunted plants well adapted to drought conditions. Mulga scrub is the dominant tree in this landscape, which is also littered with dead and dying trees. Termites are very common, providing reptiles, such as geckos and skinks, with food. These in turn are eaten by spiny-tailed goannas, mulga monitors and Gould's goannas.

Although *Caimanops amphiboluroides*, left, is an **arboreal dragon**, favouring mulga trees in particular, it may also be encountered foraging amidst leaf litter on the ground. Up to 25 cm long, it is distributed in the coastal region of Western Australia around Shark Bay and also further inland. It is fairly uncommon and most likely to be seen during daylight when it searches for insects. Generally a very slow mover, it tends to rely upon its camouflage colouring for protection against its enemies.

The **striped-tailed goanna** (*Varanus caudolineatus*), is one of Australia's pygmy monitors, growing to about 30 cm long. It is an arboreal lizard, mainly living in mulga or native cypress trees, and is widely distributed throughout the coast and interior of central Western Australia. However, because it leads a very cryptic life, spending much of its time hidden beneath bark or in hollow branches, the striped-tailed goanna is rarely seen. It is diurnal, hunting skinks, geckos and insects, and despite its elusiveness is preyed upon by birds and other enemies.

This heath country can be found along the Murchison River on the coastal plains to the south of this region. Hundreds of different plant species, which have blossomed with spectacular effect as a result of winter rains, cover the landscape which is inhabited by numerous snakes and lizards, including the thorny devil.

Found throughout the north west, the **pygmy spiny-tailed skink** (*Egernia depressa*) tends to live in family groups. Usually around 15 cm, the coloration of this skink varies with habitat. This specimen is from Mount Newman in the Ophthalmia Range east of the Hamersleys but other membersof the same species are more brightly coloured. Feral cats, foxes, birds, snakes and goannas all hunt this skink which hides in rock crevices or tree cracks. The skink then uses the spines positioned along its tail to wedge itself in. Insects and other invertebrates constitute this lizard's diet and it is generally active in the daytime.

Despite its fearsome appearance the **thorny devil** or **moloch** (*Moloch horridus*) is actually quite harmless. Its spiky exterior is intended only to deter enemies. During confrontations the moloch retracts its head under its body and presents its enemies with the spiky hump located on its neck. Up to 15 cm long, the moloch is found in a variety of habitats, from sand dune country to deserts, throughout much of Western Australia, South Australia and the interior of the Northern Territory. It is well adapted to arid environments and can soak up water with its skin, this moisture is then channelled to the mouth by capillary action. Ants constitute its sole diet.

5

THE SOUTH WEST

ROUGHLY TRIANGULAR IN shape, Australia's smallest drainage region in terms of area occupies the south west corner of the continent. It is bordered by ocean on two sides and desolate, generally waterless salt lake and sandy desert country on the other and is drained by rivers that flow variously south and west towards the Indian Ocean. Unlike the watercourses that meander for hundreds of kilometres across the Pilbara and Kimberley regions, the south west rivers are comparatively short. The three longest river systems, the Swan-Avon, Blackwood and Frankland, are 410, 300 and 240 kilometres respectively, while most, such as the 112 km long Murray, are far shorter.

All of the rivers are seasonal, carrying the largest volume of water in winter and spring, and most are also intermittent. Perennial rivers are restricted to the extreme south west where rainfall is highest and more prolonged. Most of the rivers originate in plateau country where they generally occupy fairly narrow flood plains in their upper reaches. Some, such as the Avon, Frankland and Blackwood, also run at right angles to the coast during this stage of

This swamp is typical of many similar habitats located along the southern coast and hinterland of Western Australia. It is surrounded by heath-type vegetation growing on sandy soil and acts as a habitat for several highly adapted fish species, such as Balston's perchlet and the salamander fish.

Both the **western pygmy perch** (*Edelia vittala*), right, and **Balston's perchlet** (*Nannatherina balstoni*), see page 135, are found in south western Australia's coastal swamps, although the former is far more common. These fish have been known to reach 7 cm in length but 4–5 cm is more usual. They are both active during the day, when they eat small crustaceans and insect larvae, and are related to north eastern Queensland's famous jungle perch.

their journey. The Avon flows for 320 kilometres in a south to north direction before suddenly swerving west towards the coast, bursting through the Darling Scarp and changing its name and nature to become the Swan. Many rivers are deeply incised where they break through the escarpments that surround the plateau country. Some, like the Avon which plunges through the Darling Scarp via a 36 kilometre long gorge up to 150 metres deep and a kilometre wide, occupy fairly wide valleys while others have created spectacular gorges with steep V-sided profiles. Still others have carved new courses in their original flood plains. Fan-shaped alluvial deposits spreading away from the escarpment slopes mark the point where many rivers commence their brief foray across the coastal plain. Sometimes sandbars partially block the river mouths, causing water to back up and form swampland, while sporadic lagoons and semi-permanent salt lakes indicate the location of abandoned river courses.

Physically the region may be divided into three distinct types of landform: coastal plains, scarplands and tablelands or plateau country. Although the plains stretch right around the coast in a continuous narrow belt from Broome to Albany, appearing in a more fragmented form further east before they blend in to the great Nullarbor Plain, these lowlands are most evident north of Bunbury. Except where the rivers have deposited alluvium the plains are largely sandy and frequently covered with badly drained shallow depressions, particularly in the west. These lowlands rarely exceed 36 kilometres in width. The scarplands mark the inland margin of the plains and the edge of the plateau district and run in a north-south and west-east direction throughout the region. They are most dramatically represented by the Darling Scarp, commonly called the Darling Range although this description is a misnomer as the 'range' is not a separate entity but part of the plateau itself. Stretching southwards from near Geraldton for some 360 kilometres the scarp is most impressive between Moora and Donnybrook where it rises sharply from the plain as a more or less continuous wall. Intermittently broken by rivers issuing forth from the plateau, this section of the escarpment is marked by a sharp gradient, topped by a precipitous cliff up to 30 metres high.

The plateau's southern rim is less dramatic but this area does contain several mountain ranges that rise abruptly from the level surrounding terrain. The largest and most magnificent of these is the Stirling Range which stretches for some 80 kilometres along an east-west axis 65 kilometres from the sea. It is roughly 20 kilometres wide and distinguished by a series of rugged peaks that rise up to 1110 metres above sea level. These tops are frequently enveloped in cloud and sometimes covered by winter snow. Between the Stirling Range and the sea lie the Porongorups whose rounded domes reach up to 600 metres above sea level while further east is the 460 metre high Barrow Range. The bulk of the south west region is occupied by the ancient Great Western Plateau region, known in this area as the Yilgarn Plateau, which slopes gradually downhill towards the Indian Ocean. The south west surface of the plateau is gently undulating and less fissured than the Pilbara or Kimberley districts. Although the plateau surface alternates be-

tween 300 and 450 metres in height above sea level these variations are not readily apparent because the changes occur so gradually. Much of the plateau is covered by sand-plain soils that overlie a hard layer of laterite, up to 5 metres thick and sometimes eroded by wind and rain to form breakaways. Towards the interior the plateau is dotted with salt lakes, the remnants of an ancient interior drainage system.

Climatically this region is temperate or Mediterranean, experiencing cool, wet winters and warm, dry summers. During the winter westerly airstreams accompanied by rain-bearing depressions blow across the land. When these winds move south during the summer they are replaced by dry winds that originate in the interior of the continent, producing sunny weather and clear skies. Most of the region's rain falls between May and September, with the extreme south west receiving the lion's share of between 1000–1400 mm. The western side of the Darling Scarp is also well watered along its southern coastal rim, averaging some 1200 mm per year. Further inland rainfall totals plummet drastically, averaging 400–600 mm per year, as the depressions expend most of their moisture against the western and southern edges of the escarpments or mountain ranges. These barriers create a rain shadow effect, starving their landward sides of precipitation. Temperatures are highest in January and lowest in July with the two mean maximums being 25.8 and 15.7 degrees Celsius for Albany, 29.6 and 17.3 for Perth and 30.9 and 14.5 for Narrogin further inland. Daily and seasonal extremes are greatest further inland and during the summer the western coast is cooled by a sea breeze known locally as the 'Fremantle Doctor'.

Botanically this region is probably the most varied and interesting in Australia. Many of the 6000-odd species found here are known nowhere else in the country. The region is particularly rich in heath-type plants and contains 35 out of the country's 50 or so banksia species, 112 out of 180 grevilleas, 71 out of 100 hakeas and 26 out of 30 isopogons. It also contains some of Australia's largest remaining stands of giant eucalypts, notably the majestic jarrah and karri trees which are endemic to the south west, and is renowned for its brilliant carpets of spring wildflowers, including species such as scarlet banksias, red kangaroo paws, smoke bushes and yellow mountain bells. Vegetation varies with both rainfall levels and altitude but four broad types may be discerned.

To the north the sandy and relatively arid coastal plain is covered with dense, generally low-lying scrub. Mulga often dominates but the landscape is also covered with other drought-resistant plants such as grevilleas, hakeas, casuarinas and banksias, many of which produce a dazzling flower display in spring. The well-watered south west corner, extending for some 70 kilometres to the north beyond Perth and eastwards along the southern coast to Albany, is covered by sclerophyll forest. The northern part of this zone is dominated by jarrah forest, covering a belt 320 kilometres long by 40 kilometres wide. These trees grow up to 45 metres high and despite their size thrive in poor laterite soils. Branches only appear about half way up the shaggy-barked trunks of these trees, which seldom overlap to form a continuous canopy. The southern section of this forest blends in with an

One of the most common frogs in Australia's south west region is the golden bell or **western bell frog** (*Litoria moorei*), a species that frequents vegetation in and around creeks and swamps. Sometimes seen during the day but more often encountered at night, this species hunts insects and is heavily preyed upon by snakes and birds, against which it has no known defence other than camouflage. Western bell frogs commonly grow to 8 cm long.

80 kilometre long and 30 kilometre wide tract of karri forest. Because these trees require up to seven months of moisture each year this belt corresponds with the south west region's highest rainfall area. Individual specimens of this species grow up to 90 metres high, making it one of Australia's most spectacular trees. The bottom two-thirds of the smooth-barked trunk are usually devoid of branches and both the jarrah and karri forest contain an understorey of shrubs, such as karri she oaks, peppermints, acacias, banksias and other species. Marri, tuart and other hardwood trees grow in association with the karri and jarrah forests.

As the rainfall total and its duration lessens further inland the coastal forests are replaced by parkland occupied by eucalypts such as wandoo or white gum, marri and tuart. This timber steadily thins out towards the east and the understorey is increasingly represented by various grass species. Mallee scrub dominates the fourth vegetation category and is found in the extreme south east of the region, bordering the Nullarbor Plain. This drought-resistant scrub grows up to 10 metres but is commonly much lower in stature and is well suited to sandy arid soils. The coastline is also sporadically covered by swampland, which supports a variety of grasses and sedges, while mountain areas, such as the Stirling Range, feature distinctive types of plantlife at different altitudes, with forest gradually giving way to heath.

Although there is a comparative abundance of freshwater in this region only ten species of freshwater fish have been discovered. Most of these are cold-water fish and as such are related to many species found in the south east region. The other cold-blooded animals found here are twelve frog species, including the turtle frog (Myobatrachus gouldii), twenty-five snake species and seventy-five lizard species, mainly composed of skinks.

Ranging widely throughout most of central and northern Australia, the **black-headed goanna** (*Varanus tristis*) is fond of numerous habitats, from rocky outcrops to trees. This black form is found near the Moore River north of Perth. Elsewhere this goanna is either a dark brown or light grey colour. Black-headed goannas reach 80 cm in length and eat frogs and small mammals.

The **desert banded snake** (*Simoselaps beratholdi*) favours sandy habitats throughout much of Western Australia, excluding the wet south west coastal fringe, South Australia and the interior of the Northern Territory and is particularly common in the sand dune country near Perth. Although this elapid snake is venomous its fangs are too small for it to be regarded as dangerous to humans and its secretive lifestyle and nocturnal habits ensure that it is rarely encountered. Like the ten other burrowing snakes in this genus the desert banded snake is chiefly distinguished by its brilliant markings. It grows up to 30 cm.

The karri is the second tallest tree in Australia, reaching heights of 90 m. It is endemic to the high rainfall areas of south west Australia and sometimes grows in association with another stately eucalypt, the jarrah. These magnificent white trunked trees commonly tower above an understorey of young eucalypts, ferns and grasses, a habitat favoured by various species of western reptiles and frogs.

The **salamander fish** (*Lepidogalaxias salamandroides*), top left, is an ancient species related to the pike-type fishes of the northern hemisphere. It is a bottom-feeder and spends most of its time foraging for insect larvae, worms and crustaceans. Males possess modified pelvic fins which are used during mating. This species is found in the swampy areas of southern Western Australia and averages 5 cm in length.

Three species of *Galaxias* fish live in the coastal swamps of southern Australia. The smallest and most colourful of these is the striped dwarf galaxias (*Galaxias nigrostriata*), a tiny species reaching only 4 cm in length when fully grown. It can be distinguished from the others by prominent black lines which run along the body. Slightly larger in size and less conspicuous is *Galaxias munda*. Like its two cousins it forages for small crustaceans and insect larvae. The most robust of the three is the western galaxias (*Galaxias occidentalis*), which grows up to 15 cm long and is probably the most commonly encountered species.

Usually found in dry creek beds, the **white spotted frog** (*Heleioporus albopunctatus*), right, is distributed between the Murchison and Phillips Rivers in Australia's south west region. Like several other large burrowing frogs found in this district, this species is most likely to be seen on the surface immediately prior to rain on warm nights when it leaves its hole to hunt beetles and other insects. It is one of Australia's most attractively-coloured frog species and is generally a chocolate colour, splattered with white spots. It grows up to 9 cm long.

The **moaning frog** (*Heleioporus eyrei*), named after its distinctive call, is most commonly found on the sandy plains north and south of Perth. Large numbers of moaning frogs appear during winter rain showers when they leave their burrows to hunt and breed.

One of Australia's most bizarre-looking amphibians is the **turtle frog** (*Myobatrachus gouldii*). It inhabits the sandy country in the south west of Australia where it spends most of its life hidden underground, usually near termite mounds so that it has a ready supply of food. Unlike most other burrowing frogs this species digs into the ground head-first, using its powerful front limbs. Besides being extremely odd-shaped this amphibian is also unable to swim. Its breeding habits remained a mystery until recently when it was discovered that development of the offspring took place entirely within the egg so that there is no tadpole stage. Due to its cryptic lifestyle this frog is rarely seen except occasionally during heavy rain at night.

As the landscape grows drier and more arid towards the east of this region, away from the high rainfall districts of the extreme south west and closer to the Nullarbor Plain, mallee scrub and salt bush become the dominant types of plant cover. Various species of dragons and monitors, including the ubiquitous Gould's goanna, may be found in this habitat.

The **western bearded dragon** (*Amphibolurus minor minimus*) is a semi-arboreal lizard commonly associated with arid habitats, such as mallee scrub, in Australia's south west region. Usually about **25** cm long, this dragon is equipped with excellent camouflage for its preferred environment but is sometimes seen basking in the sun on fence posts or rocks. If menaced this dragon resorts to bluff, erecting its spikey beard, although this beard is less impressive than that of its eastern cousin, the common bearded dragon.

6
THE SOUTH EAST

SWEEPING SOUTHWARDS IN A narrow coastal arc from the McPherson Range in northern New South Wales, through Victoria to the south east corner of South Australia, this region is determined by those rivers that drain from the Great Dividing Range to the sea. Tasmania is included in the region as a detached section. The largest mainland rivers, ranging from north to south, are the Richmond, Clarence, Macleay, Manning, Hunter, Hawkesbury, Shoalhaven, Snowy, Tambo, Mitchell, La Trobe, Yarra and Glenelg. Tasmania's rivers mostly originate in high country located in the centre or north east of the island and the main ones are the Derwent, South Esk-Tamar, Huon, Mersey, Arthur and Gordon.

The most obvious characteristic of the rivers that drain this region is that, with the exception of a few in western Victoria, all are perennial, although water levels generally vary throughout the year in most places. This reliability is a consequence of the region's fairly high and dependable rainfall combined with the fact that most rivers rise in high country and run rapidly towards the sea. The long intermittent and meandering watercourses common to regions such as north west Australia or the Gulf of Carpentaria district are unknown here. This combination of weather and relief can have some startling results. Tasmania, which has the smallest drainage area and some of the shortest rivers in Australia, also has the greatest run-off in the country. Many rivers, such as the Tweed or Forth, are fairly short and flow to the sea by a more or less direct route. Other rivers, such as the Hawkesbury, Manning or Hunter, are relatively long and reach the sea by a more roundabout route. The region's longest waterway, the 462 kilometre long Hunter River, initially flows along a parallel course to the coast — a preference shared by the Richmond, Clarence, Hawkesbury and Shoalhaven Rivers — and only turns eastwards after joining its major tributary, the Goulburn River.

Whatever their length the region's rivers share several similarities. Most rise in high country, either in the Great Dividing Range or the Tasmanian highlands, and commence their journey as small tumbling creeks that spill swiftly down the slopes through narrow gullies and over waterfalls, such as the spectacular 484 metre high Wollomombi Falls on the Macleay River. Once the rivers leave the steep gradients

Barely indistinguishable from its resting place, this flat or **southern leaf-tailed gecko** (*Phyllurus platurus*), opposite, clings to a lichen-encrusted sandstone outcrop. Commonly found throughout the sandstone ranges and coastal districts of central New South Wales, this lizard appears at night after spending the day ensconced in caves and crevices or beneath overhangs. It feeds on insects and is normally about 15 cm long with a large triangular head and distinctive leaf-shaped tail. It is grey-brown or grey-green in colour on the top of its body, whitish underneath and has a mottled appearance.

Alpine meadows, such as this one in the Snowy Mountains, rely upon regular rainfall to retain their verdancy. During the winter this landscape is covered by snow and even in summer, as the ripples on the water surface indicate, the temperature can be chilled by wind. This environment is the habitat of several types of high country reptiles and frogs.

South Australia
New South Wales
Victoria
Tasmania
Richmond River
Tweed R
McPherson Range
Clarence R
Doughboy Range
Wollomombi Falls
Macleay R
Wilson River
Port Macquarie
Manning R
Barrington Tops
Hunter Valley
Hunter R
Newcastle
Great Dividing Range
Hawkesbury River
Gosford
SYDNEY
Woronora R
Kangaroo Valley
Wollongong
Shoalhaven R
South Pacific Ocean
Snowy Mountains
+ Mt Kosciusko 2228 m
Snowy R
Cann River
Tambo River
Nicholson River
Mitchell R
La Trobe R
Gippsland
Yarra R
MELBOURNE
Geelong
Otway Range
Glenelg R
Tasman Sea
King Island
Bass Strait
Flinders Island
Strezlecki Range
Forth River
Tamar River
+ Mt Ossa 1617 m
Arthurs Lake
HOBART

The **corroboree frog's** (*Pseudophryne corroboree*) distinctive black and yellow coloration makes it one of Australia's best known frogs. It is found in the high country of the Australian Capital Territory and adjacent areas of New South Wales, usually above the snow line. Its favourite habitat is grassy bogs and marshland and vegetation alongside creeks. The frog's coloration is probably a form of defence mechanism as these colours are widely associated in nature with danger. Corroboree frogs are about 3 cm long.

of the high country behind they slacken in pace, becoming more sluggish so that they sometimes meander across flood plains in their middle and lower reaches. By this stage of their course most of the major rivers have had their volume boosted by extensive tributary systems. Several rivers, such as the Richmond, Clarence and Derwent, enter the sea via long and complex estuarine systems. Others, like the Hawkesbury and Cann, reach the sea by drowned river valleys while some, such as the Mitchell and Nicholson, drain into coastal lagoons and lakes.

The most obvious and imposing relief feature on the mainland is the Great Dividing Range, a broad belt of ranges, ridges and deeply dissected plateaux that extends along the east and south eastern seaboard. Despite its rather grandiose title the Great Divide rarely exceeds 1000 metres in altitude and its most important topographical function is to act as a watershed between rivers flowing towards the sea and those draining inland. The Divide reaches its highest and most impressive point in the Australian Alps, a vast plateau that straddles the New South Wales-Victoria border and which averages 1500 metres above sea level. This rugged environment includes the Snowy Mountains and Australia's highest peak, Mount Kosciusko (2228 metres). The eastward side of the Divide is sporadically marked by ranges,

such as the Barrington Tops or the Doughboy Range, that extend towards the coast, acting as barriers between the river basins. Other ranges, such as the Otway and Strzelecki Ranges, exist independently of the main divide, marooned in the coastal plain. The latter is rarely continuous and mostly exists as scattered patches, usually corresponding to river valleys or basins. It reaches its greatest extent in Gippsland and the Hunter Valley, stretching inland for 240 kilometres and 130 kilometres respectively. Elsewhere the plain is often quite narrow and is only about 16 kilometres wide at its fullest extent for most of the southern New South Wales coastline where the Great Divide swings towards the sea. The coastline itself is very diverse, featuring long sandy beaches, rugged headlands, extensive high cliff sections, large bays and drowned river valleys. Parts of the coastal hinterland are occupied by swamp and lagoons while dunes are common along much of the New South Wales coastline.

Despite its mountainous reputation only about half of Tasmania exceeds 300 metres in height while only a very small proportion of the island extends over 1200 metres above sea level. The most mountainous terrain is found in the west and north east of the island and this rugged environment has been greatly glaciated. The wild north west region contains Australia's deepest lake, Lake St Clair (220 metres), and Tasmania's highest peak, Mount Ossa (1617 metres). Tasmania's lowlands are mainly found along river valleys, such as the Derwent, that creep inland between the mountain ranges and plateaux, or else along the northern coastline.

Rainfall throughout the region is largely determined by latitude and altitude. The extreme north lies in the sub-tropical climatic zone and is affected by tropical weather systems that bring a summer rainfall maximum, although some rain falls in the winter. Elsewhere the region lies in the temperate climatic zone and is affected by a westerly airstream in winter that is associated with rain-bearing depressions. While much of Tasmania and western and southern Victoria experience a winter rainfall maximum, coastal New South Wales and east-

At 40 cm in length when fully grown the **blotched blue-tongued lizard** (*Tiliqua nigrolutea*) is one of the largest members of this family in Australia. It is a native of the highlands of Victoria, southern New South Wales and northern Tasmania and frequents open heath, woodland and montane forest, occasionally appearing on roads. Because it lives at high altitudes this species is active at much lower temperatures than its cousins. The blotched blue-tongue is a ground-dwelling diurnal lizard and it has a catholic diet, eating insects, molluscs, berries and wild flowers.

Essentially a type of **mountain trout**, *Paragalaxias mesotes* is found amongst boulders and aquatic vegetation around the edges of Tasmanian mountain lakes. Unlike other members of the same family, *Paragalaxias mesotes* is not a good swimmer and, like a gudgeon, moves in short bursts, resting for intervals in between.

ern Victoria receive rain throughout the year. Rainfall totals vary. The highest totals are recorded in the most mountainous areas, with western Tasmania receiving 3500 mm and the Snowy Mountains 3200 mm. The rest of the region receives considerably less rain although by Australian standards rainfall totals are fairly substantial and serious droughts are uncommon. Temperatures are also affected by latitude and altitude, being highest in the north and lowest in the most mountainous regions. However, there is also considerable variation between inland and coastal districts and places like Melbourne are notorious for their diurnal range. Ranging from north to south the mean maximum temperatures in degrees Celsius for January and July are, Grafton, 31.6 and 21.6; Sydney, 25.5 and 15.9; Melbourne, 25.8 and 13.3; and Hobart 21.4 and 11.5.

Throughout this region the native flora varies enormously with vegetation types ranging from sub-tropical and temperate rain forest, wet and dry sclerophyll forest, alpine cover, parkland and coastal heath and swampland. Unfortunately settlement and agriculture have destroyed the vegetation cover in some districts. Sub-tropical rain forest is found in the extreme north of the region in rugged areas like the McPherson Range that receive over 1200 mm rain per year. This type of forest is composed of two layers of broad or medium-leafed trees in which the top layer forms a continuous canopy. Typical species of tree include cedar, bloodwood, Antarctic beech, hoop pine and coachwood, many of which are festooned with hanging mosses and ferns. Smaller palms, ferns and younger trees comprise the lower level. Temperate rain forest requires about the same amount of rainfall and is found mostly in north west Tasmania. However, the forest ceiling is seldom more than 15 metres high, considerably less than that of the sub-tropical forest, and the forest interior is usually an impenetrable mass of tangled trunks and branches covered with moss and lichen. Common tree species include the Antarctic beech, Huon pine and King William pine.

Alpine vegetation can be found above 1200 metres in the Australian Alps, Tas-

Sclerophyll forest and grasslands surround this Snowy Mountains stream. Because it has reached the middle reaches of its course, the water in this stream is fairly deep and slow-moving and contains species such as the trout-cod, a relative of the Murray cod. Black snakes, water dragons and mountain tree frogs may be found along the bank.

manian highlands and in isolated areas such as the Barrington Tops. Because this environment experiences harsh weather alpine plants have to withstand high winds and extremes of temperature. In the summer when the snow melts much of the high plateau country is covered by bogs, featuring sphagnum moss and lichens, and alpine meadows filled with small exquisite flowers such as alpine blue bells, sun orchids, dwarf grevilleas, white snow anemones and trigger plants. One of the most common alpine trees is the snow gum which typically grows between 900–1500 metres. This stunted and twisted eucalypt has a smooth white trunk streaked with reds, browns and creams wherever the bark has peeled back. Antarctic beeches are also found at this altitude while the sub-alpine country lower down between 900–1500 metres is sometimes heavily timbered by trees such as the mountain ash, stringybark and candlebark.

The most commonly encountered tree cover in this region is wet or dry sclerophyll forest and both types are dominated by eucalypts. These trees usually have fairly straight trunks topped by rather sparse crowns of drab grey or blue-grey leaves, which are characteristically narrow and shiny. Wet sclerophyll forests correspond to relatively high rainfall or well watered areas and common tree species include tallow wood, blue gum, blackbutt and, in Victoria, mountain ash. Although this type of forest has a fairly continuous canopy the interior is mostly open except near creeks or gullies where a profusion of small ferns, tree ferns, epiphytes, lianas and herbaceous shrubs may be found. Dry sclerophyll forests grow in less well watered environments and although trees are generally shorter most reach 15 metres in height. They also grow further apart and the forest interior features less ground cover. Common species include ironbark, white gum, scribbly gum and spotted gum. As the terrain becomes drier, particularly in western Victoria, this type of forest gives way to parkland where the trees provide no canopy at all and where the ground cover is largely composed of grasses. Among the various types of vegetation found along the coast are heathland, covered with a variety of plants such as banksias, casuarinas, grevilleas, boronias, hakeas and callistemons, and swampland which supports grasses, reeds and paperbarks.

This **mountain trout**, opposite, once scientifically described as *Galaxias fuscus* but currently lacking a classification, is found in small creeks high above the tree line in the Victorian Alps. It needs well oxygenated fast-flowing water, with a maximum temperature of 20 degrees Celsius, if it is to survive and due to its limited distribution is fairly rare. Fortunately it seems to have few enemies, although when confined in tanks members of this species can be quite savage, frequently fighting with one another.

Galaxias tanycephalus, top, lives around the margins of Tasmanian mountain lakes, such as Lake Arthur where this one was taken, and frequents aquatic plant growth and rocks. It is a very proficient swimmer and often darts about in schools, searching for insects and their larvae. Predators such as trout can be a danger but *Galaxias tanycephalus* is sometimes able to escape detection by changing its colour, turning dark or light to suit different environments. Unlike several other species from the same family, *Galaxias tanycephalus* can survive in comparatively high water temperatures up to 28 degrees Celsius. Adults grow to 12 cm long.

White's skink (*Egernia whitii*) is distributed throughout the south eastern part of Australia, from South Australia's Eyre Peninsula to the Queensland border and across the Bass Strait to northern Tasmania. Sometimes seen basking in the sun, this live-bearing skink lives in hollow logs and rock cracks located in coastal heath, grasslands and forests. During the day it hunts insects and when approached by an enemy retreats into a log or crevice.

Thick sub tropical rain forest borders the shallow higher reaches of the Wilson River, inland from Port Macquarie on the central New South Wales coast. Despite the low water level caused by summer evaporation the river still boasts a variety of fish species, with gudgeon, eels, smelt and galaxias being found in this habitat.

Although this region has the longest history of settlement and the highest population concentrations found in Australia many local cold-blooded animal species have been able to adapt to the impact of humans and survive. The fifty or so types of fish found in the region have had to deal with several changes to their habitats, including pollution and the introduction of foreign species of fish, such as trout. The most common fish are the Galaxids, a group that includes at least twenty different species The region's sixty different species of frog include the spectacular corroborree frog (Pseudophryne corroborree) while the seventy-odd lizard species include the lace monitor (Varanus varius), bearded dragon (Pogona barbatus) and garden skink. Among the thirty-three types of snake found here are the venomous tiger snake (Notechis scutatus) and death adder (Acanthophis antarcticus) — which are both found near settled areas — and the protected diamond python (Morelia spilotes). Finally the region supports three species of tortoise.

Between September and March each year the **empire gudgeon** (*Hypseleotris compressus*), normally a rather drab yellowish-green colour, changes its appearance. Both male and female fish turn partly, and sometimes completely, red. This startling colour transformation is part of the species' breeding behaviour. Prior to spawning the male clears a flat surface on a rock or log on which the female lays thousands of tiny elongated eggs. These are guarded by the male until they hatch. Juveniles that survive the depredations of predators like water-beetles and other fish can expect to live for several years and adults sometimes reach 12 cm in length, although the average length is 8 cm. This species is distributed round the northern Australian coast, from southern New South Wales to the Murchison River in Western Australia, and lives in many habitats, from mountain streams to estuarine swamps. It is diurnal and feeds on insect larvae and small crustaceans. Wildlife found alongside the river includes diamond pythons, common green tree frogs, red-eyed tree frogs, large barred-river frogs and the tusk frog, which is often heard making a distinctive 'tock tock tock' call.

This magnificent looking **Gippsland water dragon** (*Physignathus lesueurii howittii*), opposite, has assumed its defensive role, puffing out its throat and opening its mouth ready to bite. If necessary this dragon will indeed bite an aggressor and it can inflict serious wounds. Whenever possible though this species opts for flight rather than fight. Much of its time is spent perched in tree branches above water and, if disturbed, the dragon will seek santuary by dropping straight into the water. This specimen was found near Kangaroo Valley in New South Wales; others at the southern limit of the dragon's range in Gipssland are a more greenish colour. Generally no more than 80 cm long, adults feed on small mammals and frogs as well as more pedestrian prey such as insects, and will also eat berries.

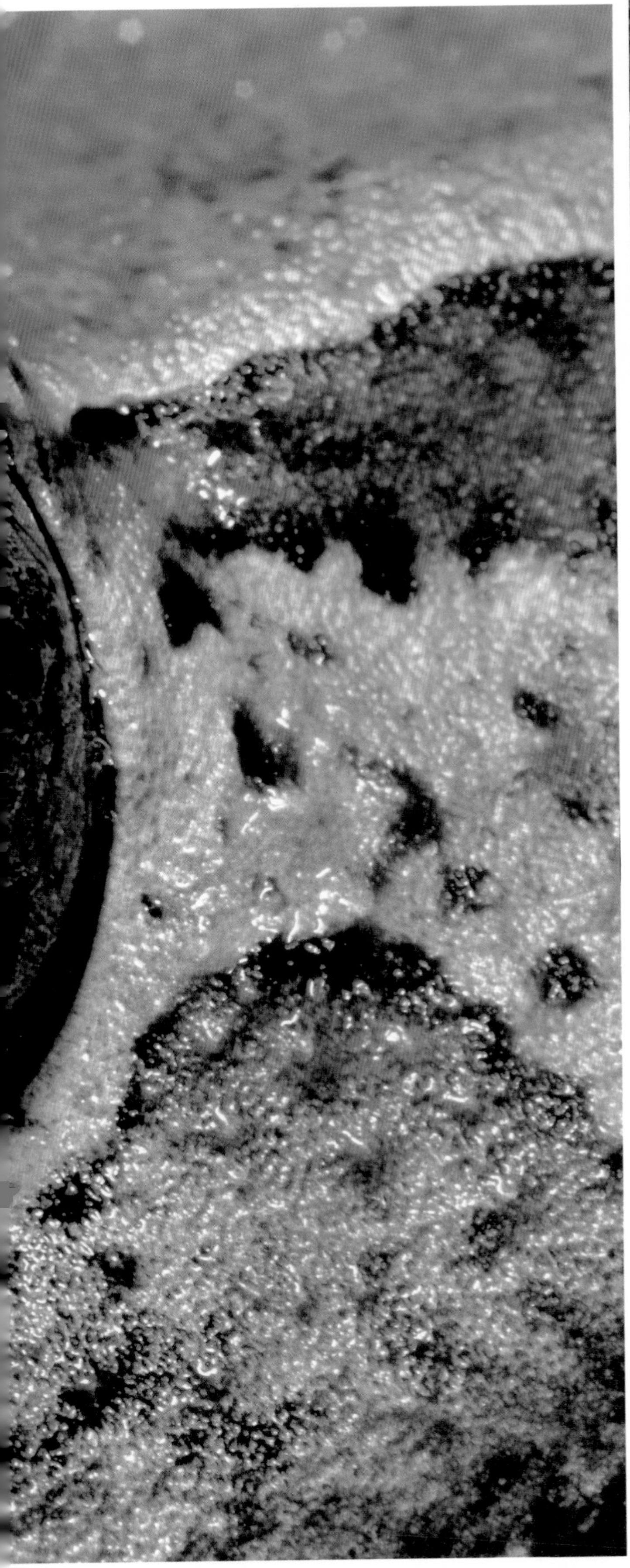

The **giant barred frog** (*Mixophyes iteratus*) grows up to 12 cm long and lives besides creeks in rain forest country between northern New South Wales and southern Queensland. Females are larger than males and during the breeding season they lay their eggs in leaf litter adjacent to streams. This strategy ensures that the eggs are washed into the water during rain, thus allowing them to hatch into tadpoles. Insects and smaller frogs are the staple diet of this species.

Like many frog species the giant barred frog has an intricate pattern of thin lines arranged over its eyes. This network probably improves the frog's camouflage while it sleeps during the day, when the eye's pupil is closed to a mere slit.

Most likely to be seen sitting on rocks but sometimes encountered in densely vegetated habitats alongside creeks, the strikingly coloured **Blue Mountain tree frog** (*Litoria citropa*), above, is distributed throughout eastern Victoria and along the New South Wales coast. The distinctive red and yellow-green splashes on this predominantly light brown amphibian are used as defensive colours, flashing warning signals to enemies such as snakes whenever the frog jumps. Up to 6.5 cm long, it is nocturnal and eats insects.

Usually found in fresh or brackish habitats throughout the coastal river systems of New South Wales, *Philipnodon spec.* is a small diurnal gudgeon about 5 cm long. It is generally observed foraging for worms and insect larvae amidst submerged tree roots and rocks.

Rain forest environments located between north coast New South Wales and the lower part of Cape York Peninsula are the favourite habitat of the **red-eyed tree frog** (*Litoria chloris*). It is also known to live in grassy river flood plains and is most likely to be seen after rain.

Keferstein's tree frog (*Litoria dentata*) is usually found in coastal swamps or lagoons between central New South Wales and southern Queensland. It is a fairly common species of frog and is often heard during the summer breeding season after rain, when males make long high-pitched wavering calls to attract mates. When these calls prove successful the frogs mate in swampy conditions. This species grows up to 5 cm long and is nocturnal, spending the night feeding on insects.

The **common galaxias** (*Galaxias maculatus*) is, as its name suggests, widely found in rivers from northern New South Wales right around the southern Australian coast to Western Australia. It is also known in New Zealand, the temperate zone of South America and several South Pacific islands. Although normally an inhabitant of fresh water, the female lays her eggs alongside river estuaries during high spring tides when water inundates the surrounding land. Each female lays several thousand eggs which are left in humid conditions amongst plants on the river bank. The arrival of the next high tide about two weeks later causes the eggs to hatch and the larvae to enter the river as the tide recedes, spending the first six or seven months of their lives at sea before swimming back up the river.

The heavy-bodied and powerful **diamond snake** (*Morelia spilotes*) does most of its hunting in trees, strangling birds, flying foxes and rodents. It is distinguished from other pythons by its distinctive gold and olive-black colouring, which is arranged in a rough diamond-shaped pattern. Still fairly common in rainforest habitats throughout coastal New South Wales, this snake grows up to 3.5 m long and has few enemies. It is a totally protected species. Females lay between ten and twenty eggs, using their coils to protect and insulate them until they hatch — a period of about three months.

Sometimes caught crossing roads at night, the **bandy bandy** (*Vermicella annulata*) is a burrowing snake and spends much of the time underground. Its striking coloration acts as a warning signal to most animals but if the bandy bandy is molested it will flatten out at the same time lifting part of its body into one or two loops. Distributed throughout all the mainland states and found everywhere except in the south west, this species has adapted to most habitats and feeds chiefly on blind snakes. It grows up to 90 cm long.

Although the venomous **pale-headed snake's** (*Hoplocephalus bitorquatus*) bite is painful, this species is not dangerous to humans. It is seen here adopting an aggressive defensive pose, rising up and throwing S-shaped loops and opening its mouth before striking. Nocturnal and partly arboreal in habit, it is found in coastal and mountain forest country between central New South Wales and southern Cape York, although its distribution is not continuous. Skinks and geckos form the major part of the snake's diet and it grows up to 50 cm long.

Often seen migrating across country in the summer, the long or **snake-necked tortoise** (*Chelodina longicollis*), opposite, lives in billabongs, swamps and slow-moving rivers along the east coast, from Victoria to central Queensland. It is also found in the Murray-Darling system. An elegant swimmer and superb hunter, the snake-necked tortoise either stalks its prey or lies in ambush beneath vegetation or mud. Molluscs, crustaceans and small fish are swallowed whole or torn apart by the tortoise's front claws. Like the other members of this genus the snake-necked tortoise has a very long neck which can be retracted sideways under its shell in times of danger. It grows up to 25 cm long and females lay ten or more eggs in summer, burying them above the water level in banks.

The common or **eastern water dragon** (*Physignathus L. lesuerii*) is a semi-aquatic lizard found living besides waterways along the east coast between northern Victoria and Cape York Peninsula. This male was found living with a harem of twelve females beside a waterfall near Gosford. Normally this dragon is very shy, plunging into the water or racing up a tree at the approach of humans, but this one stood its ground and seemed quite tame, probably having grown used to people visiting the waterfall. Among other things the eastern water dragon eats frogs and aquatic organisms and is a very able swimmer, propelling itself through the water by using its strong tail. It grows to about 90 cm in length.

Distinguished by its peculiar wrk-wrk-wrk call during the summer breeding season, the **leaf green tree frog** (*Litoria phyllochroa*) is found besides swamps, lagoons and waterways in coastal areas of southern Queensland and along the coast and hinterland of New South Wales. It grows up to 4 cm long.

One of Australia's most notorious venomous snakes is undoubtedly the **tiger snake** (*Notechis scutatus*), named after the yellow bands that sometimes cross the snake's body. It is fairly common in a number of environments, from rain to dry sclerophyll forest, throughout much of coastal and south eastern New South Wales, Victoria and south eastern South Australia. A separate species, the black tiger snake (*Notechis ater*) is found in Tasmania and south west Western Australia. Up to 1.2 m long, the tiger snake normally hunts during the day but is sometimes active on warm nights. It is particularly partial to frogs, such as this banjo frog.

For some reason this tiger snake has chosen to reverse its normal eating procedure and swallow the frog tail first. Like all snakes the tiger uses its teeth for gripping, not chewing, gradually working its way along the victim's body. To prevent suffocation while eating the snake breathes through a windpipe located at the side of the mouth. This specimen was later found to have previously swallowed six frogs.

The **Woronora River** tumbles down through rugged sandstone country south of Sydney. Similar waterways, surrounded by sclerophyll forest containing eucalypts, bottlebrushes, she-oaks and numerous other types of trees, drain from the Great Dividing Range to the central New South Wales coast. Although there is some seasonal variation in flow most rivers in this region continue to flow throughout the year, except in times of extreme drought. Numerous species of reptile, including death adders, tiger snakes, Jacky Lizards and water dragons, can be found along the river banks while the water contains smelt, galaxias, blue-eyes, gudgeon and other fish species.

The **red-crowned toadlet** (*Pseudophryne australis*) is found living in colonies in the Hawkesbury sandstone country around Sydney. Now a fully protected species, this frog is fairly rare and only found when actively looked for. It generally favours grassy habitats or detritus beside waterways but individuals may sometimes be found under rocks or logs. During the breeding period females choose moist areas on land in which to lay their eggs, which are eventually swept into water by rain. Development from the egg to the young frog stage takes about three months and adults grow to 3 cm.

Although it belongs to the most widely distributed species of goanna in Australia, this colour form of **Gould's goanna** (*Varanus gouldii*), found throughout the Sydney sandstone region, is fairly uncommon. Because of its similarity in colour to *V. rosenbergi*, a colour form found in Western Australia, there is talk of having the Sydney species reclassified. The relative scarcity of *V. gouldii* in the Sydney region has meant that little is known about its habits and breeding behaviour. It is diurnal, feeding on insects to small mammals, and known to climb trees, although it spends most of its time on the ground. It is thought to lay eggs inside termite mounds in the same way as its cousin the lace monitor (*V. varius*). This form grows up to 90 cm.

The **common death adder** (*Acanthophis antarcticus*) is the third species of this deadly genus found in Australia. This one is a reddish-brown colour and is commonly found in leaf litter, which acts as a very effective camouflage, around the Sydney region. Like all death adders it has a triangular head and narrow neck and when aroused can strike rapidly and repeatedly.

Most often seen basking in the sun on tree stumps, from which it will disappear slowly if alarmed, the **Jacky lizard** (*Amphibolurus muricatus*) is found in rocky ridges and dry sclerophyll forest throughout south east Australia. Although it prefers to run away quickly on its hind legs when frightened, this lizard will stand its ground if necessary, opening its mouth wide to disclose the yellow danger colour within. Adults grow to about 30 cm and this species feeds chiefly on insects.

The **golden perch** or yellow belly (*Macquari ambigua*) has a distinctive deep-bodied, compressed appearance and is found in the Murray-Darling river system, usually hiding behind snags in turbid water where it lurks in ambush for smaller fish and crustaceans. Golden perch only breed during floods when the water temperature is above 23 degrees Celsius. However, the creation of dams has made floods less frequent and thus reduced the frequency with which this fish reproduces. Furthermore, the flood water which is released from dams usually comes from the bottom and is generally very cold. Fortunately this species is now bred artificially. Females lay up to 650,000 eggs at a time which take between twenty-four and thirty-three hours to hatch. Adults reach 75 cm in length and weigh up to 23 kilos.

7
THE MURRAY-DARLING

EVEN BY AUSTRALIAN STANdards of scale the Murray-Darling drainage system is vast. The rivers that comprise the system flow across four states, winding through most of the interior of New South Wales and Victoria as well as a sizeable portion of southern Queensland before the Murray River reaches the sea by crossing the south east corner of South Australia. By that time the system's two main waterways, the Murray and Darling rivers, have clocked up approximately 2600 and 2700 kilometres each. The total catchment area covers just over 14 per cent of the land mass of continental Australia, draining more than one million square kilometres and dwarfing every other drainage region in the country except for Central Australia. Beside the Murray and the Darling Rivers the largest waterways in this huge system are the Murrumbidgee (1579), Lachlan (1484), Goulburn (566), Lodden (381) and Mitta Mitta (286), all of which run into the Murray; and the Macquarie-Bogan (950), Namoi (858), Condamine (690), Gwydir (668) and Castlereagh (549), which are all tributaries of the Darling. The region is bordered in the east, south and north east by the inland slopes of the Great Dividing Range and in the north by the low watersheds of the Gowan and Warrego Ranges, which barely top 350 metres above sea

This snag-filled reach of the Macquarie River, one of the New South Wales tributaries of the Darling River, is typical of the habitats found in the Murray-Darling system. These logs have either been swept down from upstream during floods or else trees have collapsed into the river as a result of currents undermining the bank. The snags are an ideal habitat for many micro-organisms which provide the basis of a food chain supporting a prolific fish population, including golden perch, Murray cod, catfish, boney bream and gudgeons. Tortoises also inhabit the water while the bank is home to tiger snakes, black snakes, brown snakes, lace goannas, many other types of reptiles and numerous frogs.

level. The western boundary is less obvious and is sporadically indicated by low ridges, reaching to between 150 and 300 metres, that separate the rivers flowing towards the Murray-Darling system and those flowing west.

In spite of the immense size of the Murray-Darling catchment area the annual discharge from these rivers is relatively small. For instance, the Darling River drains an area over thirty times that of the Clarence River on the east coast, yet the volume of water carried by the two waterways is virtually the same. The reason for this is simple; much of the Murray-Darling catchment area receives scanty and infrequent rainfall so that prolonged dry spells are common, causing many rivers to flow either intermittently or seasonally. At times, some of the rivers exist in little more than name only. The Paroo and Warrego, northern tributaries of the Darling, only reach the Darling at all during flood years and for most of the interregnum shrink into strings of pools and patches of swamp. Even the Darling itself is likely to dry up. Furthermore, for much of their length, many rivers in this system flow across generally flat countryside characterised by slight gradients. Throughout its 2700 kilometre journey the Darling descends a mere 150 metres in altitude. As a result many rivers slow to a sluggish pace with minimal velocity so that much water

The Murray cod (*Maccullochella peeli*) is Australia's largest inland fish and similar in appearance to a marine grouper. Murray cod can weigh over 100 kilos; although, due to commercial fishing, cod this large are uncommon and specimens half that size are considered rare. Found throughout the Murray-Darling system, especially in the slow-moving lower reaches, Murray cod favour turbid water conditions and often lie in deep holes or under submerged logs. They eat any fish or other creature that they can overpower and have even taken waterfowl from the surface. Cods found in the coastal rivers of northern New South Wales and southern Queensland are now considered different species.

is lost to seepage and evaporation. Once they reach their middle and lower courses many waterways begin to meander crazily across their flood plains, leaving fragmented channels, billabongs or ox-bow lakes and anabranches or distributaries in their wake. At times the labyrinth of channels and distributaries — channels which leave the main river and assume an independent course before rejoining the parent river further downstream — make it hard to discern exactly where the main channel lies. To add to the confusion some rivers change their names, usually after gathering tributaries. Hence the Condamine becomes the Balonne and then the Birree or Culgoa, depending on which channel you follow, while the Darling is preceded by the Dumaresq, Macintyre and Balonne rivers.

Many rivers rise along the inland sides of the Great Dividing Range before flowing variously west, south west and north. Initially the steep slopes and more reliable rainfall ensure that the rivers maintain a reasonable flow, particularly in the south east where the Australian Alps guarantee steep gradients and fairly heavy precipitation. This mountainous region is the source of several rivers, draining into the Murray-Darling system, including the Murray itself. The Murray winds across arid flat plains for much of its journey but the steady supply of water at its source, combined with the water supplied by tributaries originating in the same region, ensures that this waterway rarely dries up, making it Australia's largest permanent flowing river. Although the Murray's flow varies from year to year the river has only stopped flowing twice during the past century. It is far more likely to flood as a result of heavy rain, like many other rivers in the Murray-Darling system. During these occasions rivers sometimes burst their banks and inundate the surrounding lowland. Floods also intensify the normal erosion processes, causing rivers to change direction and leave newly completed billabongs behind once the floodwaters recede.

Initially the Murray hurtles down the steep gradients of the Snowy Mountains, carving deep gorges and dropping 1500 metres in only 200 kilometres. Near Albury the river begins its long meandering stage, twisting across its flood plain and fragmenting into different channels and distributaries. Until it passes Swan Hill the Murray is regularly boosted by rivers, such as the Goulburn, Campaspe and Mitta Mitta, which rise in the Victorian part of the Great Dividing Range and flow north or north west. Unlike the rivers that enter the Murray from the north, such as the Lachlan and the Murrumbidgee, or those that feed the Darling, these rivers may be relied upon to supply a large volume of water. As it continues west the Murray crosses increasingly arid terrain and is joined by the Darling River, finally running for the last 400 odd kilometres between 30 metre high cliffs to Lakes Alexandrina and Albert, from where it finally

The boney bream, also known as the **freshwater herring** (*Nematolosa erebi*), is probably Australia's most common fish and is found in freshwater habitats throughout the country. It is particularly plentiful after rain when flooding precipitates a population explosion, making this species a vital part of the food chain.

The **western carp gudgeon** (*Hypseleotris klunzingeri*) is normally found inhabiting billabongs and anabranches throughout the Murray-Darling system. It is very common but it is possible that several separate species may have erroneously been given the same name.

Although drought has greatly reduced the water level in this section of the Darling River, downstream from Bourke, the river can become a raging torrent after rain, sometimes overflowing the banks to engulf the surrounding countryside. Alternatively, stretches of the river can dry up altogether. The river red gums and black box trees growing along the banks are commonly found throughout the Murray-Darling system.

drains into the Southern Ocean.

Physically the region forms part of a vast inland plain stretching south across the continent from the Gulf of Carpentaria. As such much of the landscape, especially towards the west, is level or undulating terrain less than 150 metres above sea level. Further east the region's profile steadily ascends until it reaches the slopes of the Great Dividing Range, which is marked in places by impressive ranges or plateaus, such as the Warrumbungle or Liverpool Ranges and the Australian Alps. In the centre of the region a finger of higher ground, rarely more than 350 metres above sea level, stretches inland from the Great Dividing Range to the vicinity of Cobar, creating a natural division between the Lachlan River basin and the south eastern tributaries of the Darling. To the south west, near the mouth of the Murray River, lies an arid region of wind-blown sand ridges. Climatically, the Murray-Darling region is affected by semi-arid subtropical weather, bringing summer rain, in the north and by temperate weather, bringing winter rain, in the south. The centre is influenced by both climate zones and receives erratic and unreliable rainfall throughout the year. Generally

With the start of the spawning season the male catfish or **jewfish** (*Tandanus tandanus*) collects small rocks or pebbles inside its mouth and builds a large circular nest on the river bottom. Should a female approach the male will try to induce her to deposit her eggs in the nest. If this proves successful the male guards the eggs until they hatch. Catfish are commonly found throughout the Murray-Darling system and are most active at night when they swim along the bottom, sucking up molluscs, worms and crustaceans.

rainfall totals diminish from east to west, with the inland slopes of the Great Dividing Range and the adjacent lowlands receiving at least 500 mm per year and the extreme west receiving less than 250 mm annually. Both seasonal and diurnal temperatures can vary considerably with the greatest extremes being experienced towards the interior. The maximum means for January and July are 31 and 15 degrees Celsius.

As is normal in many settled areas, much of the native vegetation in this region has been destroyed or modified by the impact of arable and pastoral farming. However, extensive tracts still exist with different types of vegetation varying with rainfall and soil patterns. The eastern section is dominated by a wide expanse of predominantly sclerophyll woodland which extends westwards for some distance from the divide. This cover is composed of fairly large trees, up to 30 metres high, which form a fairly continuous canopy over grasses or scrub. Common species of tree include red and yellow box and red ironbark. As the landscape becomes progressively more arid towards the west the forest timber gives way to scattered timber and grasslands or scrub. The most common scrub plants are mallee, mulga and cypress pine. 'Mallee' is a general name used to describe several different species of this eucalypt, which grows up to 3 metres tall and features a dense canopy supported by slender single or multiple trunks. This type of vegetation is most common in the south west where it grows in association with heath plants, such as casuarinas and hakeas, and wildflowers on sandy soils. Cypress pines, and more commonly mulga, may be found growing further north, sometimes in association with stunted desert plants such as saltbush or bluebush. Many of the rivers are distinguished by impressive avenues of river red gums, coolibahs and black box trees. These trees sometimes reach 10 metres in height and survive periodic inundations of water when the rivers flood.

The most famous fish found in this region is unquestionably the Murray cod (Maccullochella peeli), Australia's largest true freshwater fish. Sharing its habitat are the golden perch (Macquaria ambigua) and nineteen other species, some of which have very patchy distribution. Because of the semi-arid nature of much of the region's terrain many of the thirty-two endemic frog species are burrowers, including such colourful creature as the holy cross toad (Notaden bennetti) and the salmon-striped frog (Limnodynastes salmini). The most widespread tree frog is the common green tree frog (Litoria caerulea). Reptiles found here include four species of tortoise, forty types of snakes, including nine venomous species such as the tiger snake (Pseudechis guttatus), and over 100 lizard species, of which lace monitors (Varanus varius), bearded dragons (Genus pogona) and shingle-backs (Trachydosaurus rugosus) are the most often seen specimens.

This **Macquarie tortoise** (*Emydura macquaria*) is surfacing to breathe. This species is fairly common in the Murray-Darling system and is most likely to be encountered in and around underwater snags where it seeks cover or else hunts crustaceans and molluscs or feeds on plant matter. The Macquarie tortoise breeds during summer when females lay about ten eggs, burying them high on the river bank. With the advent of winter the species hibernates. Like other members of its family the Macquarie tortoise pulls its head back under its shell in a sideways movement when threatened.

Two quite distinct colour forms of the **lace monitor** (*Varanus varius*) are known. One has broad yellow and black bands around its body and is found throughout most of the Murray-Darling region while the other has an essentially black body with yellow-orange markings and is found in much of eastern Australia. In the Murray-Darling region they live side by side and are known to mate. This Murray-Darling specimen has adopted the typical goanna defensive stance. However, after going through the usual hissing motions the goanna decided to alter its strategy by yawning, a manoeuvre which either heralds a sudden bolt for freedom or a frontal attack. These creatures are arboreal and may take to trees when startled, sometimes even climbing people if the subject is standing perfectly still. It is presumed that on these occasions the goanna mistakes the person for a tree. Females lay their eggs inside termite mounds utilising the heat inside the nests to incubate their eggs. Most adults grow to about 1.8 m long but individuals up to 2.1 m have been seen.

One of the prettiest geckos found in the Murray-Darling region is the **blotched gecko** (*Oedura monilis*). This juvenile came from the Maranoa River area in southern Queensland and frequents trees, usually living under the bark or in stumps. Other colour forms live in rocky areas and have a less blotched and more milky appearance. This species is fairly rare and little is known about its breeding and lifespan. Adults grow to about 15 cm.

One of the most common gecko species found in the northern part of the Murray-Darling region is the **house gecko** (*Gehyra australis*). Its grey-brown marbled appearance lends itself perfectly to the bark of river red gums and black box trees which commonly act as the gecko's habitat. However, this camouflage does not always prove effective against birds, snakes, goannas and other predators and when molested the gecko will readily discard its tail. After being jettisoned the tail continues to wriggle, sometimes giving the gecko an opportunity to escape.

Two species of the bearded dragon (*Genus Pogona*), opposite, are found in the Murray-Darling region. They are distinguished by their different scalations. *P. barbatus*, the species shown here, is commonly encountered throughout the region and in other parts of south east Australia, favouring habitats as varied as dry sclerophyll forest and desert country. This semi-arboreal creature is often seen sitting on fence posts, a vantage point that combines basking in the sun with searching for food. This specimen has erected its spectacular beard, a defensive tactic that sometimes leads to confusion between this species and the more famous frilled-neck lizard of northern Australia.

Although it looks fairly deep the appearance of this turbid section of the **Warrego River**, a northern tributary of the Darling River, is deceptive. The water in this intermittent river is actually only ankle-deep. Acacias and black box trees grow on the bank and extensive red sand drifts lie in the immediate vicinity. Freshwater yabbies are common in this habitat but the native fish species have been decimated by introduced carp. Numerous types of frog, including the green tree frog, the salmon-striped frog and several species of burrowing frog, also live in this environment.

The **blue-bellied black snake** (*Pseudechis guttatus*), like its cousins the king brown and the black snake, is highly venomous and liable to be dangerous if surprised. It is found in several habitats throughout the northern part of the Murray-Darling region and is most likely to be active during the day, when it hunts small mammals, frogs and reptiles. This snake's coloration varies with different environments, ranging from jet black to dark brown, and adults average 1.5 m in length.

In places the Murray-Darling system winds through predominantly arid territory, similar to this landscape near the Warrego River. Acacias, poverty bushes and Mitchell grass grow on the low sand hills of this generally flat and monotonous terrain. Despite its apparent harshness the environment is full of wildlife. Typical animals include the smooth knob-tail gecko, bearded dragon, shingle-back lizard and king brown snake as well as numerous species of skink and burrowing frogs.

This reddish form of the **common brown snake** (*Pseudonaja textilis*) comes from the Warrumbungle Range near the headwaters of the Castlereagh River, one of the eastern tributaries of the Darling. Usually this species is a light brown colour, although other forms — found throughout eastern Australia from Cape York to the Flinders Range — may be dark brown or near black. This swift-moving diurnal species is highly venomous and regarded as being very dangerous. When threatened the common brown snake raises the forefront of its body, subscribing an S-shape with the head pointed slightly downwards, and strikes wildly and repeatedly. Adults grow to about 1.5 m long and females lay between twenty and thirty eggs.

Ctenotus regius is one of the many small banded or striped types of skink found in the dry interior of Australia. This particular genus contains many similar forms which are difficult to distinguish from one another. *Ctenotus regius* is found throughout the western part of the Murray-Darling region and generally favours red desert country, salt bush and spinifex habitats. It is most likely to be spotted in the early morning or late afternoon when it chases insects and other invertebrates. Adults reach 23 cm in length.

Steindachner's gecko (*Diplodactylus steindachneri*) is found right across the northern part of the Murray-Darling region in a zone that extends into north east Queensland. As a result of this wide distribution it can be found in habitats as disparate as desert country and dry sclerophyll forest. This attractively camouflaged gecko spends most of the day hidden amongst leaf litter, secreted in rock crevices or lodged below the ground in burrows abandoned by small animals or insects. It emerges at night to look for termites and other insects and can grow up to 7.5 cm in length.

In order to survive the extreme temperatures associated with its desert habitat, the **salmon-striped frog** (*Limnodynastes salmini*) spends much of its life buried underground. This particular species of burrowing frog is found in river beds and red sand drifts. This specimen was spotted near the Warrego River. Like other members of the same family, it only emerges from its sanctuary after rain, when it mates and replenishes its body fat supply by hunting arthropods. Its bright coloration acts as a deterrent to predators but it can also elude enemies by ejecting slime and mucus which make it almost impossible to grip or hold. This species grows to about 7 cm long.

The **common water-holding frog** (*Cyclorana platycephalus*) is one of Australia's true desert frogs. Like other frogs it sheds the top layer of skin fairly regularly when it is on the surface. However, during its dormant phase beneath the ground it separates itself from the outer layer of skin before retaining it as a sheath around the body, to help retain moisture. Along with the water stored in the bladder this moisture enables the frog to live through several successive dry seasons. Because of this ability, it was highly valued by desert Aborigines as an additional water supply. After the rain the frog leaves its burrow and sits in puddles or ponds with only its eyes and nostrils visible, and adaptation which helps the frog evade predators. It grows up to 8 cm long and is found in desert habitats throughout Australia.

The rather fantastic-looking holy cross or **crucifix toad** (*Notaden bennetti*) is commonly found along black soil flood plains in the north western part of the Murray-Darling region. It is also encountered in savannah and mallee environments. The holy cross toad secretes a poisonous sticky yellow substance from its sides and front and it is most likely that the frog's bright coloration is intended as a warning to potential enemies to keep away. Like most burrowing species of frog, the holy cross toad emerges from its hole after rain and immediately starts to breed, a necessity as the breeding habitat is temporary. This species grows up to 5 cm long.

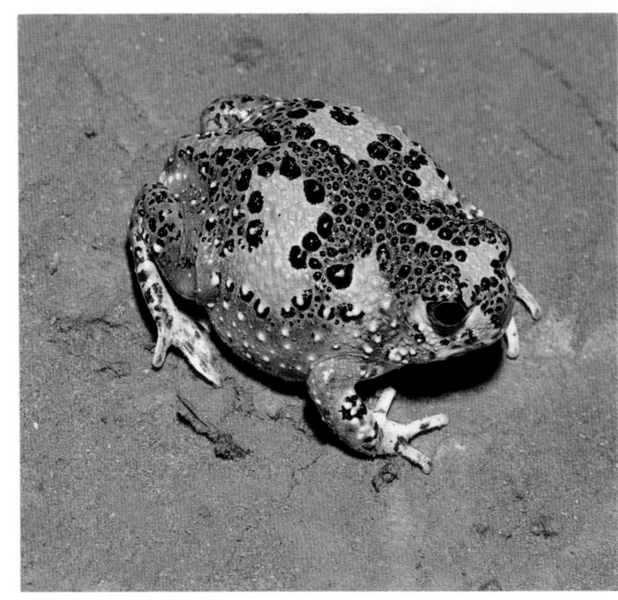

The **knob-tailed gecko** (*Nephrurus l. levis*) is widely distributed in desert regions throughout Australia, from western New South Wales to the Western Australia coast, and is invariably seen in sandy habitats. After spending the day hiding in burrows, which it either digs for itself or else inherits from small mammals and insects, the knob-tailed gecko emerges at night to look for small lizards and insects. This specimen has caught a beaded gecko (***Lucasium damaeum***). Although I have come across the aftermath of a gecko battle in the wild — indicated by converging tracks, disturbed sand and a dropped tail — I have never been able to photograph the actual feeding sequence depicted here except in captivity. This particular knob-tailed gecko was tempted with a smaller beaded gecko that had been hit, but not killed, by a car. The knob-tailed gecko stalked its victim with cat-like stealth, flicking its tail nervously from side to side. After manoeuvring itself into a striking position the knob-tailed gecko pounced, swallowing its victim tail-first — a reversal of the usual eating procedure. After finishing its meal the knob-tailed gecko licked its mouth clean of sand. This species grows up to 12 cm long and is usually a reddish-brown colour with yellow markings.

Several different colour forms of the **woma** (*Aspidites ramsayi*), a python native to the desert areas of central Australia, exist. This ground python is mostly encountered in rocky and sandy habitats and can sometimes be spotted crossing roads during the early morning or late afternoon. However, it is mainly active at night when it hunts small mammals and other reptiles. Apart from people and dingoes this snake, which grows up to 1.8 m long, has few enemies.

8
Central Australia

TWO BASIC FEATURES DIStinguish the Central Australian drainage region from its neighbours: first, with the exception of a few streams that flow into South Australia's Spencer Gulf, all the rivers in this region drain towards the interior; second, although this region covers a far greater area than any other Australian drainage region it has the lowest recorded run-off. The Central Australian region stretches right through the heart of the continent, incorporating the arid interiors of Western Australia, South Australia and the Northern Territory and spilling over into south west Queensland and north west New South Wales. Except where the Great Sandy Desert and the Nullarbor Plain confront the Indian Ocean and the Great Australian Bight respectively, the region is entirely land-locked. The drainage is dominated by the Lake Eyre catchment area which covers approximately 1.3 million square kilometres.

Lake Eyre occupies the lowest part of the continent, with the lake bed lying 14 metres below sea level, and is infrequently fed by a series of long rivers, such as the Georgina, Diamentina and Thomson, from the north and north east and by shorter rivers from the west. Lake Eyre has only been filled completely twice during the past century. Elsewhere the region is drained by rivers running south east from the low ranges that lie near the centre of the continent. Many areas, such as the Nullarbor Plain or the various sandy deserts, are completely devoid of surface drainage so that any rain that does fall merely seeps away into the ground or stands in pools briefly before evaporating. The area surrounding Lake Eyre, between the Macdonnell and Musgrave Ranges and to the extreme west is dotted with numerous salt lakes or salinas, such as Lakes Disappointment, Carnegie, Barlee and Mackay in Western Australia, Lakes Neale and Amadeus in the Northern Territory and Lakes Eyre, Torrens and Gairdner in South Australia. Many of these lakes are elongated, recalling the vanished river systems from which they originated, and all are usually dry and encrusted with salt.

All of the rivers that flow across this region are either intermittent or seasonal or both. Some, such as the Mulligan, Diamentina, Georgina and Cooper which drain from the Barkly Tableland and Isa Highlands to the north, are fed by unreliable summer rains that penetrate from the Gulf of Carpentaria. However, these rivers do not flow along their entire length — totalling 1600 kilometres for the Cooper-Barcoo system and 1125 kilometres for the Georgina for example — except in exceptionally wet years. Generally these rivers only flow along part of their upper course before most of the water seeps away into the porous soil or evaporates. The upper reaches of these rivers comprise the Channel Country, a vast network of interconnected and braided channels that flows through south west Queensland and overlaps into the adjoining states. For most of the time these rivers flow underground, making this the greatest source of artesian water in the region, but in particularly wet seasons the whole area floods, with different channels coalescing and creating wide flood plains. At other times sudden downpours can cause flash floods throughout the region, with water surging down normally dry river courses. While very few of these rivers can be relied upon to contain water at any given time, permanent sources of water are found in gorges deep in the central ranges.

Vast tracts of this region are occupied by sandy or gibber deserts which cover some 1.6 million square kilometres or 20 per cent of the entire continent. Five deserts, the Great Sandy and the Gibson Deserts to the north west, the Great Victorian Desert to the south inland from the Nullarbor Plain, the Tanami Desert to the north and the Simpson Desert to the east, almost surround the central ranges, making this the most arid and desolate landscape on the continent. The deserts are dominated by exten-

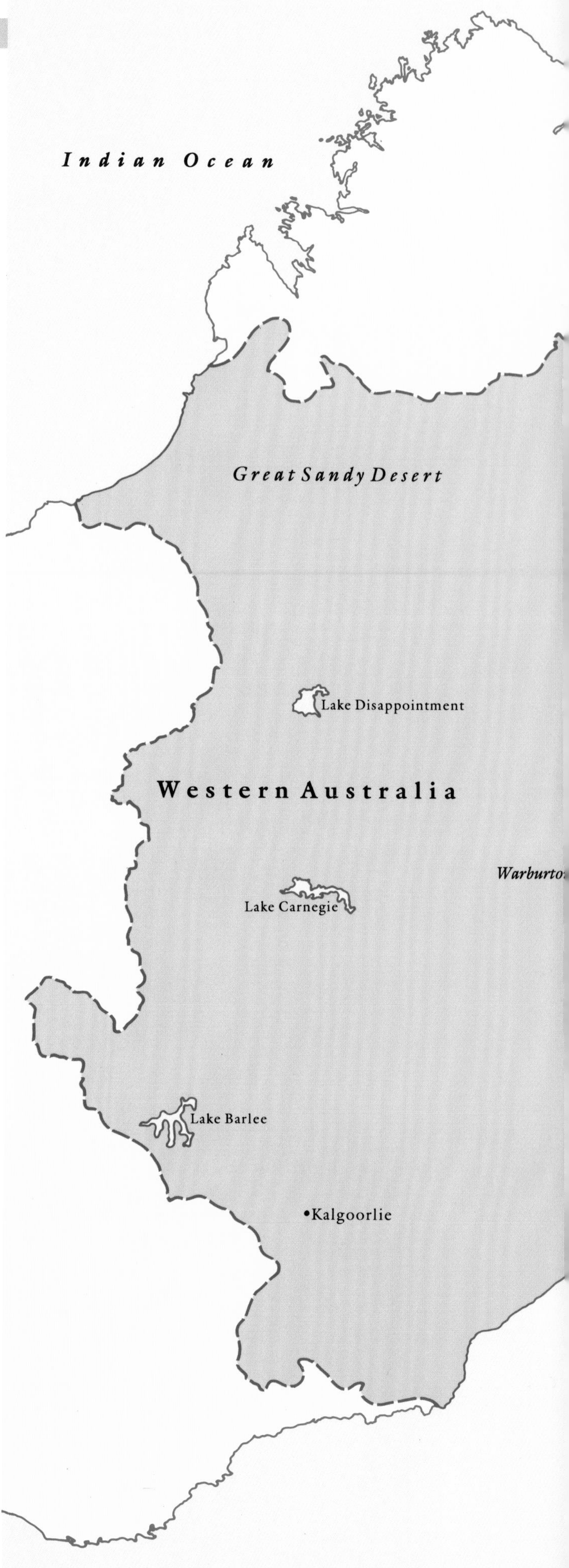

Gulf of Carpentaria
Northern Territory
Queensland
Barkly
Tableland
Tanami Desert
Tennant Creek
Murchison
Davenport
Barrow Creek
Lake Mackay
Georgina
River
Mulligan
River
Diamantina
River
Macdonnell
+ Mt Zeil 1510m
+ Mt Liebig
1524m
Alice Springs
Ranges
Standley Chasm
Ormiston
Gorge
Finke
River
James Ranges
Lake Neale
Lake Amadeus
Olgas
+ Mt Olga
1069m
Ayers Rock
wlinson
Range
Bloods Range
Petermann Ranges
Musgrave Ranges
Tomkinson Ranges
Simpson Desert
Warburton River
Thomson River
Longreach
Barcoo River
The Channel
Country
Creek
Cooper
River
Bulloo
Lake Eyre
Great Victoria Desert
South Australia
Lake
Frome
Ranges
Nullabor Plain
Lake Torrens
Lake Gairdner
Flinders
New South Wales
Whyalla
Broughton R
Eyre
Peninsula
Wakefield R
Mt
Light R
Spencer
Gulf
Lofty
Gawler R
ADELAIDE
Gulf
St Vincent
Range
Great Australian Bight
Kangaroo Island

sive sand dune systems, varying in height from 20 to 50 metres and averaging 50 kilometres in length, although dunes over 300 kilometres long have been found in the Simpson Desert. The dunes are either composed of fixed or active sand and are generally parallel to one another, although some dunes converge with tangled, maze-like results. Most dunes are fixed at their base by vegetation and only the crests are mobile, generally migrating in whatever direction the prevailing wind blows. Where the dunes are absent the deserts are commonly composed of stoney gibber, ranging from boulders to closely packed pebbles which are often polished smooth by the wind. Elsewhere red soil plains are interspersed with exposed bedrock, sometimes producing an undulating surface that resembles corrugated iron.

The ancient and greatly eroded ranges that lie astride the centre of the continent have been compared to an island in a vast sea. The endless expanses of level terrain that surround the ranges certainly help make them seem more physically impressive in stature, for they usually average a modest 300 to 600 metres in altitude and rarely stretch for more than 200 kilometres. Most of the ranges run along an east-west axis, with the major exceptions being the Murchison and Davenport Ranges that lie to the north east of the centre and the north-south running Flinders and Mount Lofty Ranges that lie along the extreme south east periphery of the region. The centre is dominated by the Macdonnells which extend for 320 kilometres west and 65 kilometres east of Alice Springs and contain the region's highest peaks, Mounts Liebig (1524 m) and Zeil (1510 m). Further south lie the James, Bloods, Rawlinson, Petermann, Warburton, Tomkinson and Musgrave Ranges. These low red rock ranges are characteristically separated by narrow valleys and contain a number of spectacular physi-

The earless dragon (*Tympanocryptis cephalus*) has a slightly misleading name, as it only looks earless. In fact, its eardrums, or tympanum, are hidden below the skin of the head. This dragon is found in dry regions between western Queensland and the Western Australia coast and is adapted to several arid habitats, from rocky slopes to gibber plains, where its coloration serves as excellent camouflage.

These table-top mountains or mesas are typical of the type of topography found in the central Australian region near **Barrow Creek**, north of Alice Springs. The surface of the mesas represents the original ground level of the landscape before the softer rocks were eroded away by wind-blown sand and water. Despite its harshness the environment is full of wildlife, including perenties, Gould's goannas, mulga monitors, woma pythons and many dragons as well as skinks, geckos and snakes.

The **mulga monitor** (*Varanus gilleni*) is one of Australia's smallest goannas, usually growing to no more than 30 cm in length. It lives in desert environments throughout central Australia and is often found on acacia or native cypress trees, usually living under bark or in hollow branches. During breeding the female digs a hole about 30 cm deep in sandy soil leading to a small chamber. After laying her eggs she re-emerges, carefully packing the top 5 cm or so of passageway to disguise her nest from predators.

cal features, such as Standley Chasm and Ormiston Gorge. Probably the most distinctive, and certainly the most famous, physical landmarks in this region are Ayers Rock, a vast monolith 867 metres high, and the Olgas, a collection of huge rounded tors dominated by Mount Olga (1069 m). These features and the nearby ranges all change colour frequently throughout the day, turning red, orange-red, purple, blue and other tones depending on the angle of the sun.

The continental climate that determines the weather in much of this region probably produces some of the most severe conditions regularly experienced in Australia. High temperatures and low, erratic rainfall are the norm. Most of the region receives less than 500 mm annually with the highest falls occurring on the northern fringe where the region is affected by tropical weather systems. Many areas receive less than 250 mm and the desolate terrain around Lake Ayre receives under 125 mm of rain. Even these meagre figures are somewhat meaningless as rainfall is generally erratic and unreliable with a succession of drought years liable to be followed by a short period of heavy rain that induces dramatic short-lived vegetation growth and sudden flooding. Given their position the central ranges receive more rain than might be expected with the higher falls being caused by altitude. Temperatures throughout the region are usually intense during the summer, frequently soaring above 40 degrees Celsius before plummeting dramatically at night due to the absence of insulating cloud cover. Winter temperatures are lower and accompanied by less extreme diurnal fluctuations. Alice Springs, located near the centre of the region, has mean maximums of 35 degrees Celsius in January and 19 degrees Celsius in July. Dust storms are common in this region with temperature fluc-

The **beaded gecko** (*Lucasium damaeum*) is commonly found living on the ground in a wide variety of dry habitats, ranging from gibber plain and spinifex-covered sand-hills to mallee and savannah country, located in the interior of all the mainland states. During the day it excavates holes or appropriates abandoned lizard and insect burrows so that it can stay under cover and evade snakes or birds. It forages at night, hunting for termites, moths and other insects. This beautifully patterned gecko is usually around 7.5 cm long and reacts to danger in the normal gecko manner, either racing away or else, when cornered, throwing off its tail as a decoy.

Terrain like this is fairly common in the centre of the Northern Territory. This vivid red rock outcrop is located just north of Tennant Creek in Australia's 'dead heart'.

tuations producing whirlwinds that sweep spirals of dust and debris up into the air for hundreds of metres.

The exceptionally harsh climatic conditions, poor soils and general lack of water endemic in this region have produced an environment where only specialised vegetation can survive. Plants commonly fall into three categories: ephemerals, perennials and species that live above ground throughout the year. The first category includes many species of wild flowers that, because they lack the essential features for survival in an arid climate, only appear briefly and quickly complete their life cycle after rain. Drought resistant seeds ensure that these plants survive to flourish fleetingly whenever sufficient rain falls. The perennials tend to die back completely when conditions become too hot and dry, only surviving because their tough root systems allow them to make a come-back when conditions become more favourable. The third category includes many well known desert plants, including the trees and scrubs that require more than a transitory

The **sandswimmer skink** (*Eremiascincus richardsoni*) is probably the most spectacular representative of the many burrowing lizards found amongst sandy or loose soil habitats throughout the drier regions of Australia. Its common name is derived from its ability to slither quickly through loose soil or sand. Due to the intense heat experienced in this environment the sandswimmer hunts at night, eating anything that it can overpower, including other lizards. This skink grows to 20 cm in length and possibly has a long lifespan as captured specimens have lived for over ten years.

The **Bredl's python** (*Morelia bredlii*) is the desert form of the carpet snake, varying only in scalation and coloration. It is found in the southern part of the Northern Territory, especially around the Finke River system where it lives in river red gums and other trees. It may also be encountered in rockier habitats where it hides from the sun during the middle of the day in cracks and burrows. Up to 2 m long, the Bredl python has few enemies except at the juvenile stage of its life and is most active at night when it hunts small mammals.

rainy spell to complete their life cycle. Eucalypts, such as the ghost or river red gums, have developed narrow elongated leaves that minimise the effects of transpiration. Other species, such as the ubiquitous mulga, shed their leaves completely once they reach adulthood and develop leaf-stalks while the saltbush and bluebush scrubs have tough leathery leaves. Although some plants, such as the mulga and many species of 'spinifex' grass, are commonly found in many desert habitats, others tend to prefer specialised niches. Hence river red gums are most likely to be found along fairly reliable watercourses, such as the ones draining from the central ranges, while ghost gums can be seen growing along hillsides. Other plants prefer the lower slopes of sand dunes. Despite the aridity of the environment very few areas exist that are entirely devoid of vegetation and after rain most areas undergo startling transformations, however brief, as plants blossom and reproduce.

Surprisingly, considering the generally arid character of the Central Australian region,

Using its powerful hind legs, this **painted frog** (*Neobatrachus centralis*) steadily twists itself into the ground. Like all but two of Australia's burrowing frog species, this amphibian buries itself 'tail'-first. This frog is so adept at burrowing that it can disappear from view entirely within a couple of minutes when excavating in loose or moist soil. The frog usually buries itself about 60 cm below the surface before settling down to await the next rains, surviving the dormant period by consuming its stored body fats. The painted frog is found throughout the drier regions of central Australia, from north western New South Wales to the Western Australian coast, and favours sand dune country and river flats. Adults grow to about 5 cm in length.

The **silvery catfish** (*Neosilurus argenteus*) is found in permanent waterholes located along the bed of the Finke River and other internal drainage systems in central Australia. It is particularly common after floods and can survive equally well in both muddy and clear water.

The **desert goby** (*Chlamydogobius eremius*), a member of the gudgeon family, is found in artesian springs and internally flowing rivers, such as the Lake Eyre drainage system, in central Australia. It favours clear shallow water, flowing over rocky bottoms and tends to live in crevices. The goby is active throughout the day and hunts crustaceans or any small fish that it can overpower. It is thought to be a survivor from the period when the present interior of Australia was inundated by sea. Coloration varies with mood and adults grow up to 5 cm long for females and 6 cm long for males.

Like other fish species found in central Australia, the **Barcoo grunter** (*Scortum barcoo*) has learnt to survive wherever it can and can be seen in artesian bores and in the permanent waterholes located along the beds of the region's drainage systems. Generally around 35 cm long, the Barcoo grunter increases its chances of survival by eating anything it can overpower.

at least sixteen species of fish are found here, inhabiting the sporadic permanent pools found along watercourses such as the Finke and Diamentina Rivers and Cooper's Creek or living in the overflows of artesian bores. However, only a few of these species are endemic and most may be found elsewhere throughout Australia. The region's most aptly named fish is the desert-goby (Chlamydogobius eremius), a species thought to have existed here when the present landscape was submerged beneath the sea. Over twenty types of frogs, mostly burrowing species that only appear on the surface after rain, live here while lizards form by far the largest category of cold blooded animals with more than 180 species being recognised. Over half of this number is composed of skinks while dragons and geckos are also plentiful. The eleven types of goanna encountered here include the perentie (Varanus giganteus) and the spiny-tail goanna (V. acanthurus brachyurus). Nine of the region's forty snake species are considered dangerous to humans and the region contains Australia's most venomous creature, the small-scaled or fear snake (Oxyuaranus microlepidota). Several small and incredibly colourful burrowing snakes may be found in sandy habitats throughout the region. A single species of tortoise is known to inhabit permanent pools in Cooper's Creek but it is yet to be properly described.

Fast-flowing, relatively deep water is best fished by using a two-man seine net, which is easily opened out by the current. Floats on the top and weights on the bottom of the net ensure that no fish escape being snared in the bag section located at the middle.

PHOTOGRAPHY

IN RECENT YEARS NATURE photography has become an increasingly popular pastime. Unfortunately though, many people mistakingly assume that photographic technique and the use of sophisticated equipment are the only ingredients required in obtaining good photographs. While these two factors are undeniably important the most crucial element in successful nature photography is undoubtedly understanding the subject and its behaviour. This vital knowledge can only be obtained through long and careful observation and is never complete. Although I started photographing animals after acquiring my first camera at the age of eighteen I am still discovering more about my chosen subject.

Naturally a degree of technical expertise and reliable equipment is necessary if one wants to be an accomplished photographer. When photographing animals my first choice in terms of basic equipment would be a single-lens reflex camera fitted with a macro lens to facilitate close focussing. The pictures in this book were taken with a Rolleiflex SL 66 camera using either an 80 mm or 150 mm lens. Ektachrome film was used for all the shots. An automatic exposure meter and a motordrive are desirable accessories as they minimise the amount of time spent worrying about the mechanics of taking a decent picture and allow the photographer time to compose the subject. Most of the diurnal pictures in this book were taken using natural light but the nocturnal and aquarium shots required the assistance of one or more flashguns. Because of the problems caused by turbid water in most natural environments the fish and tortoise specimens were all photographed in large tanks, some with a 1500 litre capacity, using up to eleven Metz electronic flashguns.

Photographing each type of creature produced different problems. The greatest difficulty in photographing fish was invariably the obtaining of decent specimens. While I collected many myself, the bulk was supplied by friends

Some of the best nature photographs only happen by chance. Situations such as this one rarely, if ever, occur in the wild. Normally the slender tree frog would have vacated its perilous position atop the green tree frog in an instant, for fear of being eaten. However, this well-fed pet green tree frog took no notice of the slender acrobat which kept jumping around on its back.

Close-up photographs of venomous snakes, like this black tiger snake, should only be attempted with the help of experienced snake handlers. It is important that at least one other person should always be watching the snake, as it is difficult to frame a picture and concentrate on the subject simultaneously.

and institutions. Before photographing the captive fish it is important that they be given time to settle into their new environment. The actual photography requires considerable patience and time and was only accomplished by exposing numerous rolls of film. Frogs are almost as hard to photograph as fish, largely because of their nocturnal lifestyle and the fact that many only emerge from hiding during or immediately after rain. Lizards and snakes also pose a challenge to the photographer as their cryptic and secretive habits often make them difficult to discover, let alone coax into the open as a subject. Ironically the wariness with which some reptiles regard humans often makes them very dramatic subjects; the frilled-neck lizard for instance only erects its spectacular frill when alarmed. Large venomous snakes should always be approached with respect and caution, especially nervous species like the taipan and brown snake. Whenever possible the subject should be photographed against an authentic background. Where this proves impossible, as in a fish tank, a suitable habitat should be created. Although photographing fish and tortoises is a solitary occupation, an expert assistant was needed to help photograph the other animals, coaxing reluctant subjects into the required posture or position or else watching dangerous species while I concentrated on framing my picture.

Direct human interference was sometimes the only way to get a good picture. Lizards, snakes and even fish were sometimes lured into the optimum position for the photographer by attracting them with food or artificial lures, such as the silver foil on a string that enticed the golden perch into the right spot. Occasionally it was necessary to interfere with the creature's habitat in a more fundamental way. By changing the water in a tank it was possible to simu-

One of the many problems inherent in photographing wild animals is getting the subject to co-operate. This large lace monitor had decided to take refuge on the drive-train of the land cruiser and could only be removed with great difficulty.

Wild freshwater environments are usually very murky or cloudy, making sharp underwater photography impossible. It is also very difficult to focus on subjects that are free to swim wherever they choose over a large area. To overcome these natural limitations all the underwater photographs in this book were taken in freshwater tanks, lighted by numerous electronic flash-guns.

The overwhelming majority of reptiles photographed in this book were encountered while they were crossing roads, a situation where their camouflage was least effective and where they were most vulnerable. This shingle back lizard was found sitting on a road in southern Queensland.

late the rainbow fish's spawning conditions and thus induce the male to metamorphose into its brilliant breeding colours. Whatever approach is used to obtain a photograph it is important to remember that the subject's welfare comes first at all times. Hence, when provoking a lizard or snake into assuming its defensive posture care should be taken not to cause the subject too much distress. Other pictures, such as feeding or mating sequences, require a thorough knowledge of the particular species' behaviour and habits. No matter what approach is used or how much one knows about the subject it is impossible to exclude the possibility that a good picture may rely largely on luck. After all, you have to be lucky to catch or see many creatures at all! But after chance has been taken into consideration all good pictures depend on experience, patience, help and reliable equipment.

Most animals will hide or run away when approached. This Spencer's goanna has had its escape routes cut off so that the photographer can obtain a good close-up shot of the subject's defensive behaviour. The goanna has complied by arching its tail ready to swing.

Unlike many birds or mammals, reptiles rarely make interesting photographic subjects unless they are made to stand their ground and react to the camera. Usually when approached they instinctively seek cover. When cornered however, they adopt various forms of defensive behaviour — sometimes with spectacular effect. This bearded dragon has erected its magnificent beard and opened its mouth in a bid to look aggresive and dangerous.

FURTHER READING

Barker, J. and Grigg, G., *Australian Frogs*, Rigby, Australia, 1977

Cann, J., *Tortoises of Australia*, Angus and Robertson, Australia, 1978

Cogger, Dr H., *Reptiles and Amphibians of Australia*, Reed, Sydney/Wellington, 1975

Gow, G., *Australia's Dangerous Snakes*, Angus and Robertson, Australia, 1982

Jenkins, R. and Bartell, R., *Reptiles of the Australian High Country*, Inkata Press, Australia, 1980

Lake, J. S., *Freshwater fishes and rivers of Australia*, Nelson, Australia, 1971
—*Australian Freshwater Fishes*, Nelson, Australia, 1978

McDowall, R. M., *Freshwater Fishes of South-Eastern Australia*, Reed, Australia, 1980

Merrick, J. R. and Schmida, G., *Australian Freshwater Fishes: Biology and Management*, Merrick, Australia, 1984

Mirtchin, P. and Davis, R., *Dangerous Snakes of Australia*, Rigby, Australia, 1982

Tyler, M., *Frogs*, Collins, Sydney/London, 1976

Tyler, M. J., Smith, L. A. and Johnstone, R. E., *Frogs of Western Australia*, W.A. Museum, Australia, 1984

Storr, G. M., Smith, L. A. and Johnstone, R. E., *Lizards of Western Australia*, W.A. Museum, Australia, 2 Vols, 1981

INDEX

Page numbers in italics indicate illustrations

Page numbers in italics indicate illustrations

Page numbers in italics indicate illustrations